Biotech & Bioethics

Editor: Tracy Biram

Volume 352

Independence Educational Publishers

First published by Independence Educational Publishers

The Studio, High Green

Great Shelford

Cambridge CB22 5EG

England

© Independence 2019

ISBN-13: 978 1 86168 808 8

Printed in Great Britain

Zenith Print Group

Contents

Introduction

BIOTECH AND BIOETHICS is Volume 352 in the *ISSUES* series. The aim of the series is to offer current, diverse information about important issues in our world, from a UK perspective.

BIOTECH AND BIOETHICS

Biotechnology and the ethical considerations it presents is an incredibly controversial topic. This book explores the many issues and innovations surrounding this fast-paced field. It also considers how recent biotech breakthroughs could shape the future for humans and the planet.

OUR SOURCES

Titles in the *ISSUES* series are designed to function as educational resource books, providing a balanced overview of a specific subject.

The information in our books is comprised of facts, articles and opinions from many different sources, including:

◆ Newspaper reports and opinion pieces

◆ Website factsheets

◆ Magazine and journal articles

◆ Statistics and surveys

◆ Government reports

◆ Literature from special interest groups.

A NOTE ON CRITICAL EVALUATION

Because the information reprinted here is from a number of different sources, readers should bear in mind the origin of the text and whether the source is likely to have a particular bias when presenting information (or when conducting their research). It is hoped that, as you read about the many aspects of the issues explored in this book, you will critically evaluate the information presented.

It is important that you decide whether you are being presented with facts or opinions. Does the writer give a biased or unbiased report? If an opinion is being expressed, do you agree with the writer? Is there potential bias to the 'facts' or statistics behind an article?

ASSIGNMENTS

In the back of this book, you will find a selection of assignments designed to help you engage with the articles you have been reading and to explore your own opinions. Some tasks will take longer than others and there is a mixture of design, writing and research-based activities that you can complete alone or in a group.

FURTHER RESEARCH

At the end of each article we have listed its source and a website that you can visit if you would like to conduct your own research. Please remember to critically evaluate any sources that you consult and consider whether the information you are viewing is accurate and unbiased.

Useful Websites

www.cam.ac.uk

www.conserve-energy-future.com

www.genomicsengland.co.uk

www.heraldscotland.com

www.independent.co.uk

www.medicalfuturist.com

www.nuffieldbioethics.org

www.pged.org

www.sciencealert.com

www.technologyreview.com

www.telegraph.co.uk

www.theconversation.com

www.thehastingscenter.org

www.weforum.org

www.yourgenome.org

www.yougov.co.uk

What is biotechnology?

Biotechnology is the use of biological systems found in organisms or the use of the living organisms themselves to make technological advances and adapt those technologies to various different fields. These include applications in various fields from agricultural practice to the medical sector. It does not only include applications in fields that involve the living, but any other field where the information obtained from the biological aspect of an organism can be applied.

Biotechnology is particularly vital when it comes to the development of miniscule and chemical tools as many of the tools biotechnology uses exist at the cellular level. In a bid to understand more regarding biotechnology, here are its types, examples and its applications.

According to the Biotechnology Innovation Organization,

'Biotechnology is technology based on biology – biotechnology harnesses cellular and biomolecular processes to develop technologies and products that help improve our lives and the health of our planet. We have used the biological processes of microorganisms for more than 6,000 years to make useful food products, such as bread and cheese, and to preserve dairy products.'

Types of biotechnology

1. Medical biotechnology

Medical biotechnology is the use of living cells and other cell materials for the purpose of bettering the health of humans. Essentially, it is used for finding cures as well as getting rid of and preventing diseases. The science involved includes the use of these tools for the purpose of research to find different or more efficient ways of maintaining human health, understanding pathogen, and understanding the human cell biology.

Here, the technique is used to produce pharmaceutical drugs as well as other chemicals to combat diseases. It involves the study of bacteria, plant and animal cells to first understand the way they function at a fundamental level.

It heavily involves the study of DNA (deoxyribonucleic acid) to get to know how to manipulate the genetic makeup of cells to increase the production of beneficial characteristics that humans might find useful, such as the production of insulin. The field usually leads to the development of new drugs and treatments, novel to the field.

Examples

Vaccines

Vaccines are chemicals that stimulate the body's immune system to better fight pathogens when they attack the body. They achieve this by inserting attenuated (weakened) versions of the disease into the body's bloodstream. This causes the body to react as if it was under attack from

the non-attenuated version of the disease. The body combats the weakened pathogens and through the process takes note of the cell structure of the pathogens and has some cells 'remember' the disease and store away the information within the body.

When the individual becomes exposed to the actual disease, the body of the individual immediately recognises it and quickly forms a defence against it since it already has some information on it. This translates to quicker healing and less time being symptomatic.

The attenuated disease pathogens are extracted using biotechnological techniques such as growing the antigenic proteins in genetically engineered crops. An example is the development of an anti-lymphoma vaccine using genetically engineered tobacco plants made to exhibit RNA (a similar chemical to DNA) from malignant (actively cancerous) B-cells.

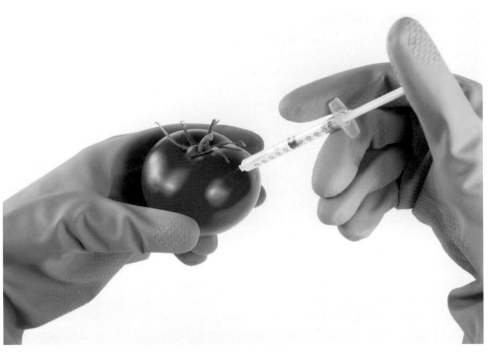

Antibiotics

Strides have been made in the development of antibiotics that combat pathogens for humans. Many plants are grown and genetically engineered to produce the antibodies. The method is more cost effective than using cells or extracting these antibodies from animals as the plants can produce these antibodies in larger quantities.

2. Agricultural biotechnology

Agricultural biotechnology focuses on developing genetically modified plants for the purpose of increasing crop yields or introducing characteristics to those plants that provide them with an advantage growing in regions that place some kind of stress factor on the plant, namely weather and pests.

In some of the cases, the practice involves scientists identifying a characteristic, finding the gene that causes it, and then putting that gene within another plant so that it gains that desirable characteristic, making it more durable or having it produce larger yields than it previously did.

Examples

Pest resistant crops

Biotechnology has provided techniques for the creation of crops that express anti-pest characteristics naturally, making them very resistant to pests, as opposed to having to keep dusting them and spraying them with pesticides. An example of this would be the fungus *Bacillus thuringiensis* genes being transferred to crops.

The reason for this is that the fungus produces a protein (Bt) which is very effective against pests such as the European corn borer. The Bt protein is the desired characteristic scientists would like the plants to have and for this reason,

they identified the gene causing Bt protein to express in the fungus and transferred it to corn. The corn then produces the protein toxin naturally, lowering the cost of production by eliminating the cost of dusting the crop with pesticide.

Plant and animal breeding

Selective breeding has been a practice humans have engaged in since farming began. The practice involves choosing the animals with the most desirable characteristics to breed with each other so that the resulting offspring would also express these traits. Desirable characteristics included larger animals, animals more resistant to disease and more domicile animals, all geared to making the process of farming as profitable as possible.

This practice has been transferred to the molecular level with the same purpose. Different traits are selected among the animals and once the genetic markers have been pointed out, animals and plants with those traits are selected and bred for those traits to be transferred. A genomic understanding of those traits is what informs the decisions on whether the desired traits will express or get lost as recessive traits which do not express.

This information provides the basis for making informed decisions enhancing the capability of the scientists to predict the expression of those genes. An example is its use in flower production where traits such as colour and smell potency are enhanced.

Applications of biotechnology

1. Nutrient supplementation

One of the biggest uses of biotechnology is the infusion of nutrients into food in situations such as aid. Therefore, it provides food with heavy nutrients that are necessary in such situations. An example of this application is the production of golden rice where the rice is infused with betacarotene. The rice has vitamin A, which the body can easily synthesise.

2. Abiotic stress resistance

There is actually very little land on Earth that is arable with some estimates placing it at around 20%. With an increase in the world's population, there is a need for the food sources available to be as effective as possible to produce as much food in as little space as possible. There is also a need to have the crops grown to be able to make use of the less arable regions of the world.

This means that there is a need to develop crops that can handle these abiotic stresses such as salinity, drought and frost from cold. In Africa and the Middle East, for instance, where the climate can be unforgiving, the practice has played a significant role in the development of crops that can withstand the prevailing harsh climates.

3. Industrial biotechnology

The industrial applications of biotechnology range from the production of cellular structures to the production of biological elements for numerous uses. Examples include the creation of new materials in the construction industry, and the manufacture of beer and wine, washing detergents, and personal care products.

4. Strength fibres

One of the materials with the strongest tensile strength is spider webs. Amongst other materials with the same cross-sectional width, spider webs can take more tensional force before breaking than even steel. This silk has created a lot of interest with the possible production of materials made from silk including body armour such as bullet-proof jackets. Silk is used because it is stronger than Kevlar (the material most commonly used to make body armour).

Biotechnological techniques have been used to pick the genes found in spiders and their infusion in goats to produce the silk proteins in their milk. With this initiative, it makes production easier as goats are much easier to handle compared to spiders and the production of silk via milk also helps make the processing and handling much easier compared to handling the actual silk strands.

5. Biofuels

One of the biggest applications of biotechnology is in the energy production sector. With fears over the dwindling oil resources in the world and their related environmental impacts, there is a need to protect the globe's future by finding alternative environmentally friendly fuel sources. Biotechnology is allowing this to happen with advances such as using corn to produce combustible fuel for running car engines. These fuels are good for the environment as they do not produce the greenhouse gases.

6. Healthcare

Biotechnology is applied in the healthcare sector in the development of pharmaceuticals that have proven problematic to produce through other conventional means because of purity concerns.

25 January 2018

What is bioethics?

Ethics is about what we ought or ought not to do. Bioethics is one branch of ethics. Since the 1970s the term has been used to refer to the study of ethical issues arising from the biological and medical sciences.

According to the *Encyclopedia of Bioethics* (1995, p. 250) it encompasses: 'the broad terrain of the moral problems of the life sciences, ordinarily taken to encompass medicine, biology, and some important aspects of the environmental, population and social sciences. The traditional domain of medical ethics would be included within this array, accompanied now by many other topics and problems.'

It is sometimes said that science moves so quickly that ethics has difficulty in keeping up. Just because something is technically possible does not mean that it should be done. It is crucial that ethical, legal and social issues raised by the introduction of a new technology are considered from an early stage. By bringing together ethical analysis and scientific understanding, society can evaluate policies and regulate developments.

The Nuffield Council on Bioethics aims to anticipate developments in medicine and biology before problems arise, providing independent and timely advice to assist policy makers and stimulate debate in bioethics.

The study of bioethics includes topics such as:

- genetic testing and screening
- reproductive and therapeutic cloning
- the use of stem cells
- embryo research
- abortion
- assisted reproduction
- prenatal screening
- end-of-life issues including euthanasia
- the use of human tissue
- organ donation

- xenotransplanation
- the use of animals in research
- genetically modified crops
- research with human subjects
- the ethics of research related to healthcare in developing countries
- patenting DNA
- pharmacogenetics
- patient confidentiality
- resource allocation.

Addressing ethical issues

There is no set method for addressing an ethical issue. However, there are some generally accepted guidelines which can be applied to an issue. As a starting point for any discussion, it is essential that information is accurate and from an objective and reliable source. It is also important to be able to distinguish between facts and opinions. Clarity of terms and expressions is crucial.

An important part of any ethical inquiry is to examine the implications of holding a particular view. Drawing up a list of the arguments on both sides, both for and against an idea, can help to focus discussion. A further step is to analyse the basis for these arguments. The conclusions of an argument must be defensible, so it is important to look for gaps, inadequacies, fallacies or unexpected outcomes. Having assessed the validity and persuasiveness of all the arguments, a decision may be reached or it may be apparent that more information is needed.

What are the common routes to a career in bioethics?

There are diverse career paths in the field of bioethics. Individuals with a wide range of backgrounds, including philosophy, science or law, may become interested in the area. Some may choose to undertake formal training in bioethics, for example by taking a Master's degree. A number of universities in the UK offer courses and training.

Genetic modification, genome editing and CRISPR

By the Personal Genetics Education Project based in the Department of Genetics at Harvard Medical School (pgEd.org). As the technology is moving at a rapid pace, pgEd recommends visiting http://pged.org/genetic-modification-genome-editing-and-crispr/ for regular updates related to this article.

Different countries and organizations define genetic modification (GM) slightly differently. In general, GM refers to making changes to a living thing's genetic information that would otherwise not occur by natural mating or reproduction. This would usually involve using methods of biotechnology, such as 'recombinant DNA,' 'gene targeting', or 'genome editing' to add, delete or otherwise change an organism's DNA. Genetic modification can also involve moving genetic material between species.

Genetically modified organisms (GMOs), including microbes, cells, plants and animals, have long been used in scientific and medical research as a way to understand processes in biology as well as the mechanisms of diseases. The use of genetic technologies to treat diseases or make other modifications in humans, called 'gene therapy,' has been attempted since the 1990s. Less than a handful of these treatments have so far been approved by safety and regulatory agencies such as the US Food and Drug Administration.

Using gene therapy to directly treat the genetic causes of diseases has long been an aspiration for physicians, scientists and patients. Some diseases, such as cystic fibrosis or sickle cell anaemia, are relatively well understood to be caused by variants in single genes. In these cases, there is hope that, if the disease-causing gene can be corrected or replaced, then it may be possible to cure individuals with the disease or at least prevent the disease from worsening. However, gene therapy is more difficult for more complex conditions such as heart disease, diabetes or many forms of cancer, which result from the interplay among many genes and between the genes and the environment.

In order to use genetic therapy to treat diseases in an individual after birth, a significant portion of cells in the relevant tissues or organs may need to be modified. This presents technical challenges to safely and effectively deliver the modification machinery and/or alternate versions of genes to the target cells, and to successfully make the changes to the cells' genome with minimal mistakes. If the modification is made

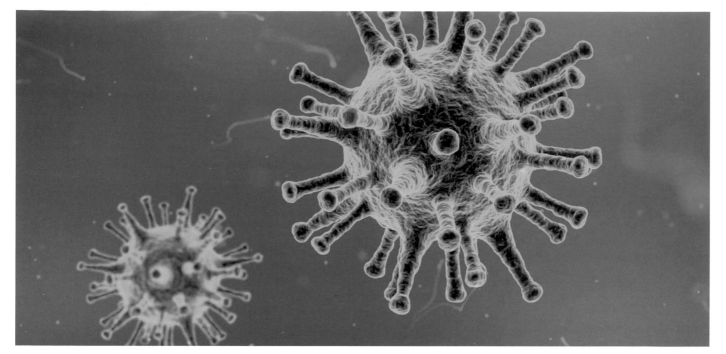

to the 'germline' (including reproductive cells as well as the cells in early stage embryos), then all cells in the body of subsequent generations will inherit that modification, as well as any mistake or unexpected change made during the process.

What is CRISPR?

Since the late 2000s, scientists began to develop techniques known as "genome (or gene) editing." Genome editing allows scientists to make changes to a specific 'target' site in the genome. One of the techniques that has generated the most excitement due to its efficiency and ease of use, is called 'CRISPR'. CRISPR stands for 'clustered regularly interspaced short palindromic repeats'. The basis of CRISPR technology is a system that bacteria evolved to protect themselves against viruses. Scientists have now taken components of the CRISPR system and fashioned it into a tool for genome editing.

There are two components to the CRISPR system: a molecule known as a 'guide RNA' (gRNA), which has the same sequence as the target site in the genome; and a 'nuclease' (i.e. a DNA-cutting molecule) called Cas9. When both of these components are delivered into a cell, the gRNA will bind to the target genomic site through complementary base pairing (meaning, As will bind to Ts and Gs will bind to Cs). In the process, the gRNA helps bring in Cas9 to the target site to make a cut to the DNA double helix. The cell's natural DNA repair mechanism will close this gap, but because the process is not perfect, a few DNA bases will be added or deleted. This renders the original gene – e.g. a gene variant linked to cancer, or one related to HIV infection – nonfunctional. Alternatively, a different version of the target gene can be placed into the cell along with the gRNA and Cas9. The cell will then use this alternate sequence as a 'template' to repair the broken DNA, copying and incorporating it into the genome. Doing so could allow an undesired version of the gene to be replaced with a desired copy.

Recent scientific breakthroughs have brought within reach the possibility of 'editing' the genome to repair disease-causing genetic variants. While it is still early days, the hope is that gene editing technologies may one day provide a cure for genetic diseases such as sickle cell anaemia, cystic fibrosis or Huntington's disease, and enable people to better fight off viral infections (e.g. HIV).

CRISPR and germline editing

Researchers have used CRISPR in cells from human, plants and animals; in fact, CRISPR has worked in all species examined to date. Notably, the CRISPR technology has been used to reverse symptoms in an adult mouse with a liver disorder and to alter DNA in non-human primates – important steps towards developing new gene therapies in humans. While genetic changes introduced into a liver cell will not be inherited in the genome of any of the individual's future offspring, DNA alterations that are introduced in the cells that will become egg or sperm, or the cells in early stage embryos, can be passed on to future generations. This is known as germline editing, and its prospects have led to discussion and debate worldwide about whether germline genetic modification in humans is appropriate, and whether or how society should proceed with such research and possible application.

On one hand, critics emphasise both the technical and ethical issues with making changes to the genome that can be passed down to offspring. There are concerns that any unforeseen effect in the editing process can become inherited. Other questions are being asked – do we have the right to alter the genome of our future generations?

Would the editing of certain diseases or disabilities lead to stigmatisation of people who are living with those conditions? And who gets to decide what are considered diseases or disabilities that should be edited? At the same time, proponents argue that germline modification can potentially eliminate diseases such as Huntington's disease, a debilitating neurological condition caused by a single gene variant. They also argue that humans have long been altering the lives and genetics of our offspring without their explicit consent, through procedures such as genetic counselling and preimplantation genetic diagnosis.

In December 2015, the US National Academies, the UK Royal Academy and the Chinese Academy of Sciences convened scientists, social scientists, ethicists and other stakeholders for an International Summit on Human Gene Editing in Washington, DC. A statement released at the end of the summit emphasized that it would be 'irresponsible' at this time to proceed with the clinical use of germline editing, but did not recommend banning the technique, instead suggesting that research should continue. In February 2017, the US National Academies' expert committee on human genome editing released its report, recommending that research on, and use of, somatic genome editing for medical treatment should continue under the existing regulatory framework, but that there should be 'broad public input' before expanding the technology's application to 'genetic enhancement'. At the same time, the report recommends that clinical trials for germline genome editing to treat 'serious diseases or disabilities' should proceed only after much more research, and only when stringent technical and ethical criteria are satisfied. Going forward, the report emphasizes the need for continued public engagement and policy debate.

Currently, germline genetic modification is illegal in many European countries and in Canada, and federal funding in the United States cannot be used for such work. As of January 2017, researchers in the UK, Sweden and China have received approval to perform gene editing in human embryos for research purposes only (in addition, existing laws or guidelines in these countries only allow research on embryos up to 14 days after fertilisation).

In November 2018, news reports emerged that the first children whose genomes were edited with CRISPR during their embryonic stage, a pair of twins, have been born in China. While the claims have still not been independently validated or published in peer-reviewed journals, the claims have drawn significant controversy. In 2019, scientists, ethicists and the broader society continue to debate the path forward.

CRISPR and the environment

CRISPR has also opened a pathway to engineer the world around us for the benefit of human health and our environment. Applications include the possibility of modifying or even eradicating disease-spreading insects, such as mosquitoes. It might also be possible to re-create long-extinct animals, such as the woolly mammoth, to roam the Earth once again, which, some scientists think, may help address climate change. However, not everyone agrees these applications would necessarily be a 'benefit', while others worry about unintended consequences of these ecosystem-changing actions.

The path forward

Gene editing has significant potential to benefit human health. At the same time, it raises profound questions that require public deliberation – what if we make alterations we regret? What if seemingly safe genetic changes prove to have unintended consequences? What are the standards for safety as the medical community seeks to explore these tools in an effort to diminish suffering? Additionally, if as a society we agree that the use of genome editing is acceptable, how do we ensure that all individuals are aware of the potentials of these technologies, and that everyone who wants to access such technologies can afford them? Researchers, bioethicists and policymakers, including a number of the scientists who pioneered CRISPR, have called for caution and the need for public consultation and dialogue that also involves faith leaders, environmental activists, and advocates for patients and for people with disabilities. As society seeks a balance between the desire to realize the benefits of gene editing and a variety of other concerns, pgEd hopes to play a part in facilitating broad conversations that engage all communities and ensure that diverse values and voices are heard.

May 2019

Read more:

• Sharon Begley, 'No red line against CRISPR'ing early embryos, experts rule' (*STAT*, February 2017)

• David Cyranoski, 'CRISPR gene-editing tested in a person for the first time' (*Nature*, November 2016)

• Antonio Regalado, 'Meet the Moralist Policing Gene Drives, a Technology That Messes with Evolution' (*MIT Technology Review*, June 2016)

• Erica Check Hayden, 'Should you edit your children's genes?' (*Nature*, February 2016)

• Ed Yong, 'What Can You Actually Do With Your Fancy Gene-Editing Technology?' (*The Atlantic*, December 2015)

• Carl Zimmer, 'Editing of Pig DNA May Lead To More Organs For People' (*New York Times*, October 2015)

• Nathaniel Comfort, 'Can We Cure Genetic Diseases Without Slipping Into Eugenics?' (*The Nation*, July 2015)

• Andrew Pollack, 'A Powerful New Way to Edit DNA' (*New York Times*, March 2014)

Related lesson plan:

pged.org/lesson-plans/#CRISPR

The synthetic biology revolution is now – here's what that means

An article from **The Conversation.**

THE CONVERSATION

By Claudia Vickers, Director, Synthetic Biology Future Science Platform, CSIRO

We live in an era where biotechnology, information technology, manufacturing and automation all come together to form a capability called synthetic biology.

Technological revolutions are significant because they shape the future of social and cultural development – as is evident for the industrial revolution, the 'green revolution', and the information technology revolution.

What is synthetic biology?

Synthetic biology is the design and construction of new, standardised biological parts and devices, and getting them to do useful things.

Parts are encoded using DNA and assembled either in a test tube or in living cells – and then applied to deliver many different kinds of outcomes.

'Cell factories' for production of industrial chemicals is one way synthetic biology is applied.

The chemical butanediol is used to make 2.5 million tonnes of plastics and other polymers each year, including half-a-million tonnes of Spandex (Lycra). In 2011, all of this molecule came from petrochemicals. Biotech and chemical companies Genomatica and BASF collaborated to engineer a commercially viable synthetic biology production route for butanediol – it went from lab to commercial scale in just five years.

Many other global businesses are also investing heavily in the use of whole cells – so-called chassis cells – to produce useful chemicals.

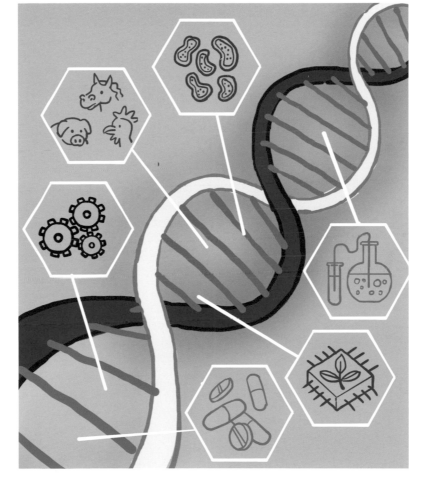

Medicine, the environment and agriculture

Significant medical breakthroughs are happening via synthetic biology.

The antimalarial treatment artimisinin can now be produced by yeast, avoiding the need to isolate it from Chinese sweet wormwood plant. This helps to stabilise global prices.

In 2016, a new immune cell engineering treatment resulted in a 50% complete remission rate in terminally ill blood cancer patients, with a 36% remission rate achieved in a 2017 trial. A similar approach has been used just recently to cure an advanced breast cancer.

Biomonitoring is another exciting area for synthetic biology developments. Highly specific, tiny biosensors can be engineered to detect an enormous range of molecules – such as hydrocarbon pollutants, sugars, heavy metals and antibiotics.

These can be applied to measure aspects of health, and in environmental sensing systems to identify contaminants.

Synthetic biology also has agricultural applications. It can provide more precision and sophistication than earlier gene technologies to help increase crop and livestock yields, while reducing environmental impact by limiting the use of chemicals and fertilisers. More efficient plant use of water and nutrients, photosynthetic performance, nitrogen fixation and better resistance to pests and diseases are all being developed using synthetic biology.

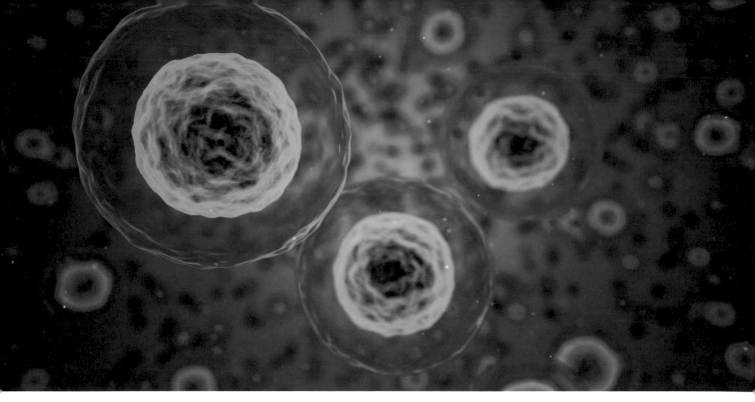

Consumer benefits may include nutritional improvements, enhanced flavour and the removal of allergenic proteins from milk, eggs and nuts.

Most of these synthetic biology applications rely on altering, adding or deleting gene functions by targeted genetic modifications. Based on past consumer resistance to genetically modified food products, progress in this area is more likely to be limited by the degree of public acceptance than it is by the technological possibilities.

Synthetic biology also provides the opportunity to use agricultural production systems for cheap, large-scale production of products such as drugs and antibodies for medical treatments.

On the up and up

International growth in synthetic biology is remarkable. In 2015, the synthetic biology component market (DNA parts) was worth $US5.5 billion – by 2020, it will approach $US40 billion. Those figures don't count sales revenue from synthetic biology products.

Product markets are also growing dramatically. In 2008, bio-based chemicals were only 2% of the US$1.2-trillion-dollar global chemical market. In 2025, that will rise to 22%, driven by development of synthetic microbial factories.

Government investment into synthetic biology has been very strong over recent years. Road maps and associated development structures have been developed through public agencies in many advanced economies, including the US, UK, EU, China, Singapore and Finland.

Private investment in synthetic biology is also growing at a remarkable rate. According to the US-based synthetic biology advocacy organisation Synbiobeta, American synbio companies raised around US$200 million in investment in 2009. In 2017 it rose to US$1.8 billion and as of July 2018 it was already US$1.5 billion, with a projected 2018 investment of just over US$3 billion.

Australia is catching up

In Australia, synthetic biology is less developed – but things are moving fast.

In 2014, the professional society Synthetic Biology Australasia formed, and several specialist synthetic biology conferences and workshops have been held.

In 2016, CSIRO invested A$13 million into the CSIRO Synthetic Biology Future Science Platform (SynBioFSP). Internal reporting shows SynBioFSP is now a A$40-million research and development portfolio driven by a collaborative community of over 200 scientists from CSIRO and over 40 national and international partner organisations, contributing to 60 research projects.

Synthetic biology was recognised as a priority area in the 2016 National Research Infrastructure Roadmap. A special call for synthetic biology was made in 2017 and a steering committee to examine Australia's synthetic biology infrastructure needs has recently been created.

This week the Australian Council of Learned Academies released *Synthetic Biology in Australia: An Outlook to 2030* as part of its horizon scanning series. We are two of the authors on this report, which examines the opportunities and challenges for getting the most out of synthetic biology in the Australian context.

Synthetic biology is an extremely fast-moving technology with extraordinarily diverse applications. It offers massive potential for Australia in terms of developing new markets, and in future-proofing in the long term.

5 September 2018

Most Brits say gene editing to reduce risk of disease should be allowed

One in ten support changing a child's intelligence or appearance through gene editing – with young people most likely to consider it for their own future children.

By Victoria Waldersee

Gene editing is a form of technology by which scientists can add, remove or alter a living organism's DNA.

The technology has the potential to radically reduce, if not eliminate, the risk of genetically inherited conditions such as cystic fibrosis or Huntington's disease. A form of gene editing called CRISPR is already being used by scientists in China to alter disease-causing genetic mutations in human embryos.

However, some are concerned about the ethical implications of other potential uses of gene editing – including changing DNA to alter a person's appearance or personality.

New YouGov research explores the extent to which the public supports gene editing, and whether they would consider using the technology to alter the genetic makeup of any future offspring.

Editing genes to reduce the risk of disease should be allowed, say majority of Brits

The British public largely support using gene editing to prevent people from passing on hereditary genetic disorders. Three-quarters (76%) of the adult population think it should be allowed, with just one in eleven (9%) saying it should be not be allowed. One in six (14%) don't know.

Older generations are slightly more likely than younger people to support gene editing to reduce the risk of disease, with eight in ten (80 – 82%) of those aged 50 and above in favour compared to 74% among 18-to-24-year-olds.

Younger people and men more likely to support gene editing for intelligence and appearance

Brits are much more reluctant to support gene editing to boost either the brains or beauty of their future children. Seven in ten (71%) oppose editing genes for intelligence, and three quarters (76%) are against doing it for appearance.

However, one in eight (12%) people are in favour of allowing gene editing for intelligence and one in 12 (8%) for appearance. Men are significantly more likely to support the process in both instances, with one in six (17%) thinking it should be allowed to change genes affecting intelligence (compared to 7% of women), and one in nine (11%) support it to alter appearances (compared to 5% of women).

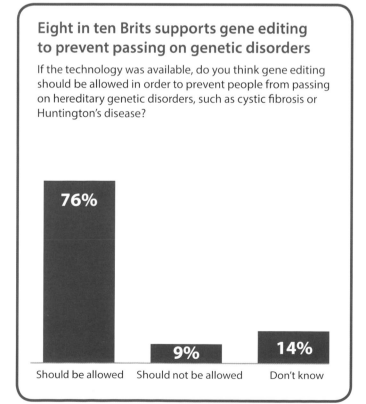

Eight in ten Brits supports gene editing to prevent passing on genetic disorders

If the technology was available, do you think gene editing should be allowed in order to prevent people from passing on hereditary genetic disorders, such as cystic fibrosis or Huntington's disease?

76% Should be allowed
9% Should not be allowed
14% Don't know

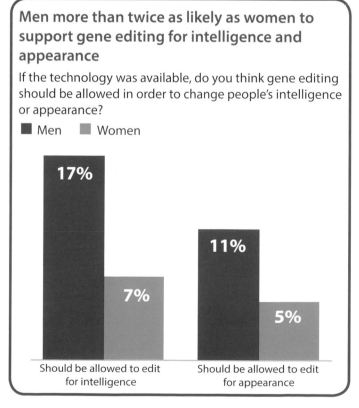

Men more than twice as likely as women to support gene editing for intelligence and appearance

If the technology was available, do you think gene editing should be allowed in order to change people's intelligence or appearance?

■ Men ■ Women

17% / 7% Should be allowed to edit for intelligence
11% / 5% Should be allowed to edit for appearance

Younger people are also more likely than older generations to support gene editing for both intelligence and appearance. Around one in six (15%) 18-to-24-year-olds support editing genes affecting intelligence and one in eight (13%) for appearance. By contrast, just one in 11 (9%) of those aged 65 and above support changing genes for intelligence, and one in 20 (5%) for appearance.

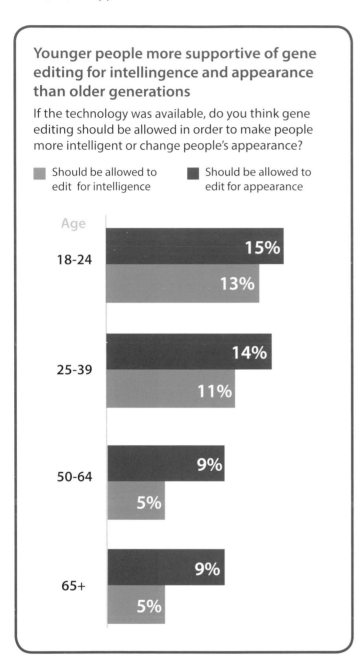

Younger people more supportive of gene editing for intellingence and appearance than older generations

If the technology was available, do you think gene editing should be allowed in order to make people more intelligent or change people's appearance?

■ Should be allowed to edit for intelligence   ■ Should be allowed to edit for appearance

Age

18-24
15%
13%

25-39
14%
11%

50-64
9%
5%

65+
9%
5%

A third of those who support gene editing for appearance think editing skin colour should be allowed

Which if any, of the following aspects of someone's appearance do you think it should be permissable to change through gene editing? Please tick all that apply *(asked only to people who said gene editing for appearance should be allowed)*

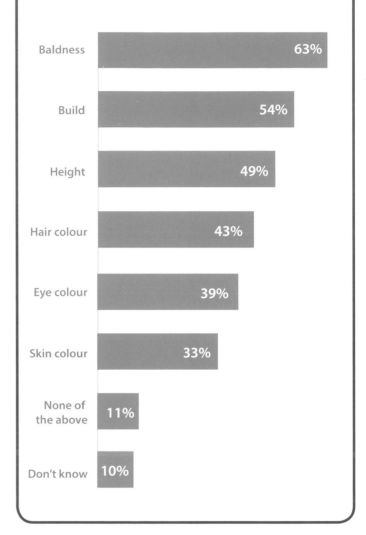

Baldness 63%
Build 54%
Height 49%
Hair colour 43%
Eye colour 39%
Skin colour 33%
None of the above 11%
Don't know 10%

How far should we go?

Though 8% of the population – equivalent to approximately four million Brits – support gene editing to alter appearance, among this group opinions differ on what aspects of appearance it should be permissible to alter.

The only characteristics which at least half of this group agree are worth editing are baldness (63%) and build (54%). Just under half think it should be permissible to edit genes for height (49%), while slightly fewer favour altering DNA to select hair colour (43%) and eye colour (39%). A third (33%) of those who favour editing genes to alter appearance say it should be permissible to edit genes for skin colour.

Would you edit your baby's genes

We also asked people who were thinking about having children how they would feel about gene editing their own offspring, were the technology available. The vast majority (83%) would consider editing if they were carrying genes for a disorder that could be passed on to their offspring (just one in nine (11%) wouldn't consider it).

Despite the fact that just one in ten Brits think gene editing to change someone's intelligence should be allowed, around a quarter (23%) would consider it to change someone's intelligence, rising to three in ten (30%) prospective fathers.

However, changing genes to alter their own child's appearance is a less popular idea, with 12% of people thinking about having kids saying they would consider it. Again, men are more likely than women to consider it for their children (15% of men vs 8% of women).

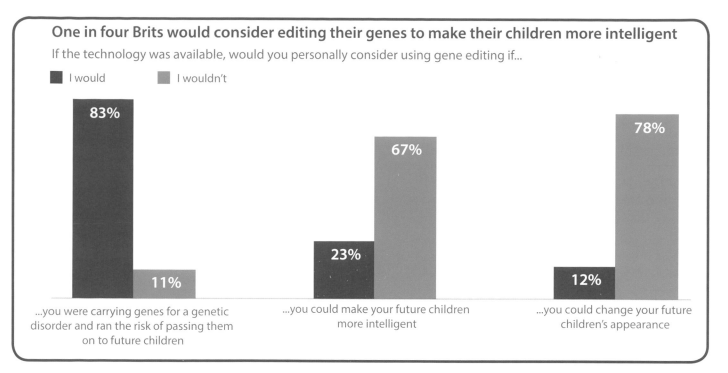

One in four Brits would consider editing their genes to make their children more intelligent

If the technology was available, would you personally consider using gene editing if...

■ I would   ■ I wouldn't

- 83% / 11% — ...you were carrying genes for a genetic disorder and ran the risk of passing them on to future children
- 23% / 67% — ...you could make your future children more intelligent
- 12% / 78% — ...you could change your future children's appearance

Approaching half of Brits (47%) believe that gene editing for particular traits is similar to choosing a sperm or egg donor based on ethnicity, build, intelligence, etc., while one in three think it is different. One in five (21%) don't know.

The journey to 100,000 genomes

Genomic potential

Pinpointing the beginning of the 100,000 Genomes Project isn't easy. It could be argued that Crick, Franklin and Watson started it all in 1953; or Frederick Sanger's pioneering sequencing technologies in the late '70s; perhaps the Human Genome Project in 2003; or even the UK10K project in 2008.

Our journey, however, began in 2012 with the announcement of the Project and, in 2013, the creation of Genomics England to drive it to completion.

The background to the odyssey was a recognition that advances in genomics, informatics and analytics brought closer the possibility of more precise diagnosis, alongside personalised and targeted treatments. In 2012, science could see the potential to identify the underlying cause of disease, predict how a person might respond to specific interventions and determine who was at risk of developing an illness.

The stumbling block? Nobody had ever tried.

The UK was brave enough to lead the way – announcing the ground-breaking 100,000 Genomes Project, which aimed to sequence 100,000 whole genomes from around 70,000 participants with rare disease, their families and people with some cancers. The decision was backed by robust government support – both political and financial – which included over £300 million of investment.

Building a genomic infrastructure – partnership on an industrial scale

As we celebrate the sequencing of the 100,000th whole genome just five years later, it is important to remember the scale of the achievement. For those of us here at the beginning, the prospect was simultaneously exciting and daunting. We were asked to not only sequence an unprecedented number of whole human genomes, but also to plug this in to the rich health data held by the NHS. With important objectives to create a consent-based and transparent programme that fostered an emerging UK genomics industry – our ultimate aim was to bring real and lasting change to NHS care.

The challenge, described by one colleague as 'building the plane whilst flying it', was to create the infrastructure for genomic medicine from scratch, whilst also delivering on our objectives. A more traditional model would have seen us build the machinery first and then begin sequencing: the 100,000 Genomes Project did both simultaneously. It was a p roject that demanded partnership on an industrial scale – and would never have been delivered without the support of thousands of organisations and individuals.

Genomics England began its close relationship with NHS England to recruit the first participants in 2014. In the same year, NHS England created 11 Genomic Medicine Centres (GMCs), eventually growing to 13 in 2015, and today joined by organisations in Northern Ireland, Scotland and Wales.

Building a genomic infrastructure

RECRUITMENT: recruited over 70,000 rare disease and cancer project participants to achieve the 100,000 genomes target by December 2018.

GENOMIC HUBS: established 13 NHS England Genomic Medicine Centres (GMCs) – alongside operations in Northern Ireland, Scotland and Wales – to recruit patients, take samples and feedback results.

INDUSTRY: brought government, academia, researchers and industry together in the 'Discovery Forum' – to accelerate translation of genomic discoveries into the clinic and catalyse economic growth.

SEQUENCING: created one of the world's largest Next Generation Sequencing Centres with our sequencing partner, Illumina – delivering the lowest price and latest technology.

STORAGE: built a multi-petabyte datacentre storing the highest fidelity whole genome DNA sequences with participant's longitudinal clinical data de-identified format.

BIOINFORMATICS: developed one of the world's few semi-automatic bioinformatics pipelines – transforming genomics from a 'cottage industry' to one suited to a health system scale.

ENGAGEMENT: embedded participant experience at the heart of decision making, engaged with the public through the 'Genomics Conversation' and built an ethical and transparent consent framework.

RESEARCH: formed the Genomics England Clinical Interpretation Partnership (GeCIP) – a global network of >2,700 researchers grouped into 42 specialist 'domains' – improving interpretation.

GMCs work across areas of three to five million people in over 85 NHS trusts. They have been key components of the project: recruiting and consenting patients; providing DNA samples; developing the mechanisms for validating results; and working to feed back results to participants. We cannot underestimate the commitment, skill and hard work of thousands of NHS staff who have pioneered these GMCs and the cause of genomic medicine in the NHS.

At the start of 2015, another piece of the infrastructure puzzle fell into place with the opening of the NIHR Biosample Centre to store samples from the Project. And just two months later, we established the Genomics England Clinical Interpretation Partnership (GeCIP). GeCiP brings together thousands of researchers and clinicians from across the world – granting them carefully controlled access to our database to power new discoveries in genomic medicine. Today GeCIP research covers 42 areas – known as 'domains' – including rare cardiovascular and neurological disorders, and cancers such as breast, lung and ovarian.

Early in the Project, we realised that many of the technologies and services needed to deliver genomic medicine simply didn't exist – and if we wanted them we would have to build

them. One of these technologies was the bioinformatics pipeline, which is critical to ensure processing at the scale in the world's largest publicly funded health system. It has involved a huge commitment from our bioinformatics team and others – but has resulted in one of the world's few semi-automatic bioinformatics pipelines.

This 'if you can't buy it, build it' approach has seen a range of innovations. Work with our sequencing partner, Illumina, for example, has acted as a significant catalyst in reducing the costs of sequencing: from billions in 2003 to around £600 today and around £100 in the not-too-distant future. We have also created a bespoke, multi-petabyte storage environment to cope with the grand challenge of managing a large-scale whole genome and clinical dataset.

In 2015 we also began to explore how to align our work with the needs of industry – the companies that will eventually turn genomic discovery into routine treatments. This started with the GENE Consortium, which evolved in 2017 into the Genomics England Discovery Forum we see today. The Forum is a platform that allows us to bring together charities, patients, researchers, clinicians and others with industry partners to share perspectives and better understand how to speed discoveries from the laboratory bench to the patient's bedside. In the past few months, the value of the Forum has been demonstrated with members discovering previously undiagnosed patients in our database – with the hope that this will lead to better diagnosis and the development of more effective treatments.

People-powered healthcare

From the outset, the Project recognised that genomic medicine could not succeed without the understanding, trust, acceptance and consent of patients. Genomic medicine is truly 'people-powered healthcare'. It heralds a changing relationship between the patient and the NHS, with a new consent model where healthcare and research become indivisible. It is important to understand that new technology and the use of data will only be socially and clinically enabled if it is trusted by the patient.

In 2016, Genomics England established the Participant Panel, which acts as an advisory body to our Board. The Panel is at the heart of our decision-making processes, with members sitting on the Access Review Committee, the Ethics Advisory Committee and the GeCIP Board. Involving participants at this fundamental level ensures that the Project is always responsible to the people who drive it.

It also underlines the importance of the Project to participants. Participant Panel Deputy Chair, Rebecca Middleton, has said, 'The Project brings me something new – hope. Whether in five or 15 years, new genomic discoveries may be able to help me.'

Ensuring that we are able to gain the trust of patients demands that we understand their attitudes to genomic medicine – and this inspired the 'Genomics Conversation' in 2016. The Conversation is a genuine engagement project – not seeking to influence its audiences, but rather to listen.

And what is it that patients get from the investment of their trust, understanding and consent?

Even at this early stage, genomic medicine is helping to transform cancer services – making real progress in providing DNA of sufficient quality for whole genome sequencing – an issue that has hampered efforts to apply genomics in cancer diagnosis and treatment around the world. NHS England is re-aligning its laboratory services as it moves from formalin-fixed paraffin embedded (FFPE) to Fresh Frozen (FF) sample handling.

The first participants received their diagnoses in February 2015, when we had sequenced around 2,000 whole genomes, with the first diagnosis of children made in January 2016.

As we have moved to 100,000 genomes, the number of people has grown. People like Project participant Alexander and his family who in finally receiving a diagnosis for LEOPARD Syndrome can now know what is wrong and seek support from others living with the same disease. Fellow participant, Jessica, and her family discovered that her condition is caused by errors in the SLC2A1 gene that cause 'Glut1-deficiency syndrome' – which only affects around 500 people worldwide. Jessica's diagnosis has opened up the potential for highly tailored treatments.

Beyond 100,000 genomes

The 100,000 Genomes Project has been a real innovation – from bioinformatics to computing to storage to research to industry partnership to public engagement. As our learning and dataset grows, so too will our ability to better diagnose and treat an ever-expanding number of diseases. Whilst the UK is now an acknowledged leader in population genomics, this is a truly global effort – with the potential to bring patient and economic benefit across the world.

The Secretary of State of Health and Social Care's announcement on 2 October laid out an exciting road map for genomic medicine. His words demonstrate the importance of personalised medicine and its ability to continue to deliver cutting-edge care within the NHS: 'I'm proud to announce we are expanding our 100,000 Genomes Project so that one million whole genomes will now be sequenced by the NHS and the UK Biobank. I'm incredibly excited about the potential for this type of technology to improve the diagnosis and treatment for patients to help people live longer, healthier lives – a vital part of our long-term plan for the NHS.'

The lasting legacy of the 100,000 Genomes Project is the NHS Genomics Medicine Service that began to roll out at the beginning of October 2018. The Project has proven the concept of genomic medicine at scale and built the infrastructure that underpins the GMS. Genomics and the GMS are transformative, poised to change the way future healthcare is delivered. As we reach the 100,000 genome milestone, we are at a tipping point in medicine that will usher in an era of highly personalised medicine – which may consign generic drugs and treatments to medical history.

Getting to this point has been a demonstration of the power of partnership. We are only here because of the thousands of patients and their families who have placed their trust in a project at the cutting edge of healthcare. Working with them have been many thousands of NHS staff who worked tirelessly to not only deliver the Project, but in many cases, pioneer totally new systems, processes and procedures to ensure that genomic medicine can become part of routine NHS care. And beyond them there is the vast and rich ecosystem of charities, industry partners, funders, government organisations and a host of others. To everyone – we would like to say 'thank you'.

Whatever its history and whenever it began, we can be sure that the genomic medicine journey is just beginning – and its future will be an exciting one.

6 December 2018

www.genomicsengland.co.uk

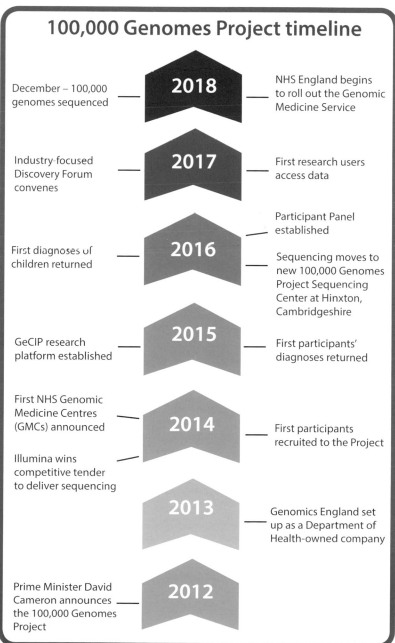

100,000 Genomes Project timeline

2018
December – 100,000 genomes sequenced — NHS England begins to roll out the Genomic Medicine Service

2017
Industry-focused Discovery Forum convenes — First research users access data

2016
— Participant Panel established
First diagnoses of children returned — Sequencing moves to new 100,000 Genomes Project Sequencing Center at Hinxton, Cambridgeshire

2015
GeCIP research platform established — First participants' diagnoses returned

2014
First NHS Genomic Medicine Centres (GMCs) announced — First participants recruited to the Project
Illumina wins competitive tender to deliver sequencing

2013
Genomics England set up as a Department of Health-owned company

2012
Prime Minister David Cameron announces the 100,000 Genomes Project

YouTube, persuasion and genetically engineered children

An article from **The Conversation.**

By George Estreich, Instructor, Oregon State University

THE CONVERSATION

On Sunday, 25 November, the scientist He Jiankui claimed the birth of the world's first genetically engineered children: twins, created by IVF, their DNA altered at fertilisation. Changes like these, because they're inheritable – 'editing the germline' – are widely prohibited by law and avoided by scientific consensus. If He really did this, it's a very big step across a very bright line.

Also, He announced the feat in a YouTube video.

The strangeness of this choice cannot be overstated. Ground-breaking achievements normally appear in prestigious journals, with extensive data, after rigorous peer review. Announcing the accomplishment on YouTube is the social media equivalent of walking out the front door and yelling, 'Guess what, everybody? I'm the first to engineer a human being! And the kids are already here – they're twins!' The timing of the video's release – on the eve of a major international conference on genome editing, where He was scheduled to speak – clearly had more to do with publicity than science.

Others have written on the science and ethics involved. I'm a writer, so what interests me is persuasion: the way literary tools, like story and metaphor, help pave the way for cutting-edge biotechnology. Researching a forthcoming book on this topic, I came to see that human-focused biotech and persuasion form a single system: for the biotech to be adopted, the public has to accept it first. He's video is a textbook bid for acceptance, an argument for germline editing aimed at the general public.

But the video is a low-rent production. It's just He, standing in a lab, talking to the camera in English (it's subtitled). As such, the video doesn't hold a candle to the polished emanations of established research institutes or multi-million dollar corporations. The writing isn't great either. Effective persuasion guides us lightly from one place to another: the lightest touch on your shoulder, redirecting your path. Listening to He is more like being yanked down a slippery slope.

And yet for precisely that reason, the video deserves a closer look: seeing the pitch in its most obvious form, we can learn to recognise the patterns.

The family success story

The video begins with He beaming at the camera, describing 'two beautiful little Chinese girls named Lulu and Nana', a few weeks old, 'as healthy as any other babies'. According to He, their parents, Mark and Grace, had always wanted to have a family. But Mark is HIV-positive, and stigma had deterred them. Now, because of He's experiment – which was intended to make the twins permanently immune to HIV infection – a happy family exists. 'The babies are home now with their mom Grace and their dad Mark', He says.

'Mark' and 'Grace', their real names changed for privacy, may be real. Rhetorically, though, their function is to humanise a new technology. Ironically, that technology is changing what it means to be human in the first place.

Portrait of a scientist

Persuasion means crafting a persona, and He is clearly going for approachable scientist and family man. Identifying himself as 'a father of two girls', He also suggests his own humility, by saying, 'Mark's words taught me something I didn't fully appreciate'. That lesson? 'Gene surgery' helps more than a child: 'We heal a whole family.'

That feigned humility doesn't square with what He actually did: rushing to be first across the germline, thereby putting his experimental human subjects, and any of their descendants, at risk. Given all this, He's assertion that he's willing to brave controversy on behalf of the parents – 'I'm willing to take the criticism for them' – rings just a tad hollow.

Metaphors, omissions and weasel words

Obviously, selling a new technology means putting it in the best possible light. Trying to do this, He makes an odd choice. Instead of 'CRISPR-Cas9', the common name, he insists on the phrase 'gene surgery'. It's a naked attempt to make CRISPR sound precise, like a molecular scalpel. That metaphor is misleading. CRISPR is improving in accuracy, but as geneticist Eric Topol wrote in *The New York Times*, unintended edits still occur – 'We don't have the assurance yet that CRISPR provides laser-like precision in editing – and we might not always detect them. Our ability to discern these changes is still rudimentary, and it is entirely likely that we will miss something', Topol added.

But the surgery metaphor has another function: It likens the radically new to the comfortably familiar. He continues this theme with a second (mixed) metaphor: 'The surgery removed the doorway through which HIV enters to infect people.'

A doorway is easy to imagine – and who could oppose slamming the door on AIDS? And yet the metaphor glosses over the complications of actual biology. It is true that people with a variation in the CCR5 gene have natural resistance to HIV. But other strains can infect the body via a different protein, which He left untouched. There is, in other words, more than one doorway. Complicating things further, having a variant CCR5 gene may protect you from HIV, but it also makes you more vulnerable to dying from West Nile virus or the flu.

Also omitted from He's video: the number of failed attempts required to get one successfully engineered baby. He's team began with 22 embryos, but in the end only one pregnancy succeeded. Of the implanted twins, one, at best, is protected against HIV. (Our genes come in pairs; in one twin, only one gene was modified, not both.)

So when He asserts that 'the surgery worked safely, as intended,' you have to remember that 'safe', like 'health' and 'choice', is a useful weasel word: positive but vague, its meaning dependent on context. 'Safe' could simply mean that the babies were born and appear to be okay. It does not mean they will be free from unanticipated effects, or protected from HIV.

Reason, emotion and the dismissal of critics

In arguments for new biotechnologies, it's common to deride critics as fearful or irrational. He's video is no exception. In it, he asserts that 'the media hyped panic about Louise Brown's birth as the first IVF baby'. Building on this theme in a second video (there are five in all), He inveighs against the phrase 'designer baby', contrasting 'vocal critics' with silently suffering families. By implication, you're either pro-technology or pro-suffering. That's a false binary, of course: untested treatments can lead to suffering.

But He's pitch also illuminates a common problem. In arguments like these, the categories of reason and emotion are invoked in contradictory ways. If people disagree with you, they're dismissed as panicky and irrational. If they're sympathetic parents who bolster your case, though, then their emotions are authoritative. In the second video, He contrasts parents and an unnamed naysayer:

'[The parents] may not be the director of an ethics center quoted by *The New York Times*, but they are no less authorities on what's right and wrong, because it's their lives on the line'.

Actually, it's their children's lives on the line. Also, invoking sympathetic parents is itself an emotional appeal.

Nana, Lulu: You've got mail

Because independent verification is still lacking, reports about Lulu and Nana tend to use the phrase 'if true.' It's appropriate, somehow, that the possibility of human germline enhancement – insistently discussed and envisioned, from bioethics conferences to movies with genetically upgraded superheroes – still seems half-imaginary, a projection, an event with deep roots in the digital and shallow roots in the real.

At the end of the video, He invites you to email his lab and share your thoughts. But weirdly – and, by the end of the video, the bar for weird is very high – you can email Lulu and Nana themselves at DearLuluandNana@gmail.com. Perhaps they'll have Twitter and Instagram accounts soon? Perhaps, one day, there'll be a LuluAndNana.com, with an online fan club and emailed testimonials? In the meantime, the twins, engineered or not, expand into the world of information, where persuasion tries to reproduce itself. To go, in our biology-based metaphor, viral.

The twins' email address is, of course, a PR gimmick. Emailing two Chinese infants (and why don't they have their own addresses? They're two people, after all) is about as meaningful as texting in a vote for this year's *American Idol*. But we can learn something from the ploy: that inheritable human modification is too serious a matter for fake participation, and that a more substantive engagement is called for. If our species is to be engineered, then we all ought to have a say.

3 December 2018

3D 'organ on a chip' could accelerate search for new disease treatments

Researchers have developed a three-dimensional 'organ on a chip' which enables real-time continuous monitoring of cells, and could be used to develop new treatments for disease while reducing the number of animals used in research.

By Julia Bradshaw

The device, which incorporates cells inside a 3D transistor made from a soft sponge-like material inspired by native tissue structure, gives scientists the ability to study cells and tissues in new ways. By enabling cells to grow in three dimensions, the device more accurately mimics the way that cells grow in the body.

The researchers, led by the University of Cambridge, say their device could be modified to generate multiple types of organs – a liver on a chip or a heart on a chip, for example – ultimately leading to a body on a chip which would simulate how various treatments affect the body as whole. Their results are reported in the journal *Science Advances*.

Traditionally, biological studies were (and still are) done in Petri dishes, where specific types of cells are grown on a flat surface. While many of the medical advances made since the 1950s, including the polio vaccine, have originated in Petri dishes, these two-dimensional environments do not accurately represent the native three-dimensional environments of human cells, and can, in fact, lead to misleading information and failures of drugs in clinical trials.

'Two-dimensional cell models have served the scientific community well, but we now need to move to three-dimensional cell models in order to develop the next generation of therapies,' said Dr Róisín Owens from Cambridge's Department of Chemical Engineering and Biotechnology, and the study's senior author.

'Three-dimensional cell cultures can help us identify new treatments and know which ones to avoid if we can accurately monitor them,' said Dr Charalampos Pitsalidis, a postdoctoral researcher in the Department of Chemical Engineering and Biotechnology, and the study's first author.

Now, 3D cell and tissue cultures are an emerging field of biomedical research, enabling scientists to study the physiology of human organs and tissues in ways that have not been possible before. However, while these 3D cultures can be generated, technology that accurately assesses their functionality in real time has not been well-developed.

'The majority of the cells in our body communicate with each other by electrical signals, so in order to monitor cell cultures in the lab, we need to attach electrodes to them,' said Dr Owens. 'However, electrodes are pretty clunky and difficult to attach to cell cultures, so we decided to turn the whole thing on its head and put the cells inside the electrode.'

The device which Dr Owens and her colleagues developed is based on a 'scaffold' of a conducting polymer sponge, configured into an electrochemical transistor. The cells are grown within the scaffold and the entire device is then placed inside a plastic tube through which the necessary nutrients for the cells can flow. The use of the soft, sponge electrode instead of a traditional rigid metal electrode provides a more natural environment for cells and is key to the success of organ-on-chip technology in predicting the response of an organ to different stimuli.

Other organ on a chip devices need to be completely taken apart in order to monitor the function of the cells, but since the Cambridge-led design allows for real-time continuous monitoring, it is possible to carry out longer-term experiments on the effects of various diseases and potential treatments.

'With this system, we can monitor the growth of the tissue, and its health in response to external drugs or toxins,' said Pitsalidis. "Apart from toxicology testing, we can also induce a particular disease in the tissue, and study the key mechanisms involved in that disease or discover the right treatments.'

The researchers plan to use their device to develop a 'gut on a chip' and attach it to a 'brain on a chip' in order to study the relationship between the gut microbiome and brain function as part of the IMBIBE project, funded by the European Research Council.

The researchers have filed a patent for the device in France.

3 October 2018

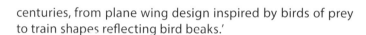

Could an Australian bee solve the world's plastic crisis?

By Jonathan Pearlman

Researchers believe an Australian bee which produces a 'cellophane-like' material for its nests could help to end the world's reliance on disposable plastics.

The native *Hylaeus nubilosus* masked bee, known for the distinctive yellow badge on its back, does not sting or live in hives but it has generated interest because of the nesting material it produces, which is non-toxic, waterproof, flame-resistant and able to withstand heat.

A biotech company in New Zealand, Humble Bee, is trying to reverse-engineer the material in the hope of mass producing it as an alternative to plastic.

Veronica Harwood-Stevenson, the firm's founder, said she began investigating the potential plastic alternative after noticing a throwaway line in a research paper about the 'cellophane-like' qualities of the masked bee's nesting material.

'Plastic particles and chemicals have permeated ecosystems and organisms around the world, [from] foetal blood of babies [to] the most remote arctic lakes; it's so pervasive, it's terrifying,' she told *The Sydney Morning Herald*.

'It's about biomimicry, about copying what's in the natural environment, and we've been doing it in design for centuries, from plane wing design inspired by birds of prey to train shapes reflecting bird beaks.'

Richard Furneaux, a chemistry professor at the Victoria University of Wellington, said the discovery of the new material was 'almost too good to be true'.

'Its robustness is beyond what you would have expected,' he told the Thomson Reuters Foundation.

Scientists analysed the genetic makeup of the bioplastic by studying the bee's glands.

Humble Bee plans to initially use the material to make outdoor apparel, such as camping gear, which often use toxic chemicals to keep them waterproof.

'Outdoor apparel is definitely what we're most interested in because of the chemicals being used and because chances are, if you like the environment, you don't want the products you enjoy to be screwing up the environment,' Ms Harwood-Stevenson said.

Scientists believe chemicals used to change the properties of plastic – such as those that make it harder or waterproof – may be harmful and could increase the risk of heart disease, cancer or infertility.

The bioplastic could also be used for aviation, electrics and construction products. It is resistant to acid which could allow it to coat medicines and help them to pass through the stomach.

The firm hopes to start selling the bioplastic in five years.

18 August 2018

Five children have new ears grown from their own cells in a world first

By David Nield

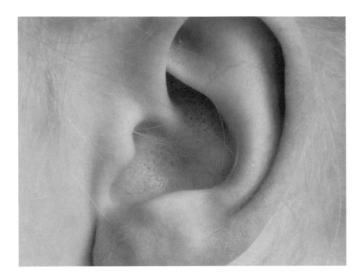

A group of five children in China have been given new ears – based on detailed 3D models and grown from their own cells – in a world first for this kind of treatment.

The kids, aged between six and nine, all had microtia, where the external part of the ear ends up deformed. In these cases the condition was unilateral, affecting only one side, so scientists were able to create high-resolution scans of their healthy ears to help grow replacement ones.

Now the team of tissue engineers and plastic surgeons has proved these techniques can work in human beings, they could offer a new lease of life for people living with microtia or other similar conditions.

'The results represent a significant breakthrough in clinical translation of tissue engineered human ear-shaped cartilage given the established in vitro engineering technique and suitable surgical procedure,' write the researchers in their published paper.

Cartilage cells called chondrocytes were harvested from the non-deformed ears by the scientists and then used to create a new ear for the other side of the head through a process of cell culturing.

With the help of computed tomography or CT scans of the healthy ears, a 3D-printed framework was created that the newly growing ear could expand into. Over time, natural cells replaced almost all of the artificial scaffolding.

Finally, the new ears were attached and reconstruction was completed, with some small cosmetic surgery procedures applied afterwards.

This kind of biological technology is actually several years old, but this is the first time it's been used so effectively in human beings – the first of these implants was fitted 30 months ago, suggesting the long-term prospects are good.

'The delivery of shaped cartilage for the reconstruction of microtia has been a goal of the tissue engineering community for more than two decades,' Lawrence Bonassar, a biomedical engineering professor from Cornell University in New York who wasn't involved in the study, told Jacqueline Howard at CNN.

'This work clearly shows tissue engineering approaches for reconstruction of the ear and other cartilaginous tissues will become a clinical reality very soon. The aesthetics of the tissue produced are on par with what can be expected of the best clinical procedures at the present time.'

Microtia rates are as high as 17.4 in 10,000 in some countries, affecting both hearing and self-confidence of the kids who are born with it.

Current treatments include fitting an artificial ear frame, or creating a new ear from rib cartilage, but this new approach beats them all in terms of both appearance and lessening the damage on the patient's body.

'Surgeons have been toying with the idea of removing cartilage tissue from a patient and distilling that tissue into individual cellular components and then expanding those cellular components,' Tessa Hadlock from the Massachusetts Eye and Ear Infirmary, who wasn't involved in the study, told CNN.

'The thing that is novel about this is that for the first time, they have done it in a series of five patients, and they have good long-term follow-up that shows the results of the ears that were grown from that harvested cartilage.'

However, there are some caveats to note – 2.5 years is a good stretch but the artificial parts of the ear haven't yet fully degraded, so further monitoring up to five years is going to be needed before we're sure this is a success.

What's more, two of the cases showed slight distortions in the growth of the ears, which scientists will have to carefully monitor.

Nevertheless it's a promising step forward for these procedures, as well as a potentially life-changing new option for those with microtia, if it becomes widely available.

'We were able to successfully design, fabricate, and regenerate patient-specific external ears,' write the researchers. "Further efforts remain necessary to eventually translate this prototype work into routine clinical practices.'

31 January 2018

Synthetic egg mayo on your human burger? The future of food

By Vicky Allan

The future of food, some people think, belongs to laboratories and factories, not farms. In giant vats, muscle and fat cells will be cultured then made into so-called 'clean meat', without, if the technology progresses, a single animal having to die. Wine – with an almost biblical flourish – will be produced not from grapes, but from, water, ethanol, sugar and other substances. Eggs will be made from plants.

This may all sound like sci-fi, but in fact the basic technology is already here. Much of the pioneering of this 'cellular agriculture' has taken place in the tech labs of Silicon Valley. It's already possible to buy milk that is a vegan simulacrum of the real thing. The lab grown burger, first developed in the Netherlands, has been around for half a decade.

Such meat has been described as 'clean meat' since it doesn't come with the same welfare and ethical implications that meat from farmed animals does. Those that promote it see it not just as answering an ethical issue, but as providing a solution to the environmental impact of our meat consumption. It has been estimated that by 2030, globally, the average human will be consuming just under 45kg per year (that's the size of a child) – 10% more than today – and that the impact on the planet will be simply too great.

In a recent survey it was found that 60% of vegans would be happy to eat lab-grown meat. However, the rest of the public are not quite so enthusiastic. Only 18% of UK respondents said they would be up for eating it. This hi-tech food is also a major disruptor of traditional agriculture. Just like the death of the CD thanks to streaming music, or the chaos on the high street thanks to online retailers, we need to ask what will happen to farmers – and the downstream companies like distributors who rely on them – if these Frankenfoods become commonplace. And, as you will discover as you read on, not everyone sees this new technology as ethical, or the real answer to our environmental problems.

Whether you're ready for them or not, here are some of the hi-tech foods coming to a plate near you, either now, or sometime in the near future.

Lab-grown meat

The technology to produce lab-grown meat has been with us for a while. The first lab-grown hamburger, was cooked and eaten in London, in 2013, the creation of Professor Mark Post of Maastricht University. It was assembled from 100,000 small strips of muscle that were individually grown at his lab.

There was a problem, though. That single burger actually cost £220,000 to make. Since then, Post has managed to tweak the process to bring it down to a cost around £4,400 per patty. But there is a significant barrier to reducing the costs further. The source of protein for the production of the meat is a serum made from animal foetal blood, which is both enormously expensive, and also, not entirely cruelty free – and hardly a sales pitch for vegetarians.

Since then, others, in the US, have developed their own cultured meat. Memphis Meats, for instance, who have created the first lab-grown meatball, say they hope to have clean duck or chicken meat on the market by 2021. The company JUST, formerly called Hampton Creek, has made a chicken using cells from a feather. They describe, on their website, being able to eat meat that comes from a bird that is still running around alive. But all these meat industry disruptors face the same problem – the issue of finding a replacement for the animal-based 'starter' serum. That's the big technology they are all trying to develop – a growth medium that is serum-free and won't frighten the vegetarians.

Plant meat

The race has been on for some years now to create a plant-substitute for meat that mimics the real thing. At the heart of this, has been the faux meat burger, pioneered, in the US, by two companies, Beyond Meat and Impossible Foods. But there are such patties in development in the UK too. Last year, Moving Mountains, said it had created the 'closest replica yet' to animal meat in the UK from a recipe that includes coconut oil, wheat and soy, potatoes, mushrooms and beetroot juice to make it 'bleed'. Company founder Simeon Van Der Molen observed that their patty 'requires less land, less water and produces less greenhouse emissions' than real meat.

Not everyone, however, is convinced this is the future. Investigative food journalist and *Sunday Herald* restaurant reviewer, Joanna Blythman is a critic. Analysing the ingredients in the Impossible Burger, she said: 'It's the very antithesis of local food with a transparent provenance and back story. I'd have absolutely no chance of tracing the origins or uncovering any substantive detail on the assiduously guarded production methods behind its utterly anonymous components.'

Synthetic wines

Last year global wine production fell to its lowest level since the 1960s because of extreme weather conditions – and experts have said this is a trend that is only likely to continue due to climate change. The answer, some scientists believe, could lie in synthetic grape-free wine. Ava Winery, a San Francisco start-up, has already made such 'wines without vines' – they literally turn water into wine, by adding ethanol, acids, amino acids, sugars and organic compounds. Its first was a sweet, sparkling replica Moscato d'Asti.

Non-dairy milk, cheese and eggs

You don't have to look to the future for a vegan milk that has been manufactured to taste just like the real thing – it is already in shops in the US now, produced by a company called Perfect Day. The process involves putting cow DNA into yeast so that it produces the milk proteins, then fermenting them with corn syrup. Replica cheese is also already on the market, as well as synthetic egg products.

The human steak

In the future we could eat ourselves. That, say some, is a possibility that lab-grown meat brings – that we could use our own cells, or those of a friend or a family member, to create a culture of meat. Evolutionary biologist Richard Dawkins tweeted last month: 'I've long been looking forward to this. What if human meat is grown? Could we overcome our taboo against cannibalism?' Owen Schaefer, a professor at the Centre for Biomedical Ethics at the University of Singapore, has even predicted that we may soon read trend pieces, saying 'Kids are eating their friends now!' It's just a shame we couldn't take this technology and use it to make chocolate. If we could it would take the expression 'if you were chocolate you'd eat yourself' to a whole different level.

22 April 2018

Biotech minnow Angle triumphs with cancer blood testing kit

By Julia Bradshaw

Biotech company Angle has scored a win after its blood testing technology was central in helping doctors discover a potentially ground-breaking treatment for cancer.

The Aim-listed firm's device, called Parsortix, is able to catch tumour cells circulating in the blood through a simple blood test.

Now researchers at Basel University Hospital in Switzerland have shown that Parsortix can also harvest tumour cells that are attached to each other in clusters, called circulating tumour cell clusters, or CTC clusters. These are highly metastatic, meaning they spread the cancer easily, but research has been limited because until now it has not been possible to harvest them.

'Where these clusters are present in the bloodstream the patient outcome is very poor,' said Andrew Newland, chief executive of Angle.

'What Basel wanted to do was investigate these clusters and it was only by using Parsortix that they could harvest them.'

The researchers found that it is these CTC clusters, rather than individual tumour cells, that are largely responsible for spreading cancer to other parts of the body. But rather than trying to kill them through chemotherapy or other treatments, which have side effects and can cause cancers to gain resistance to treatment, the scientists attempted to break the clusters using drugs that are already on the market, but not intended for cancer. They screened roughly 2,500 of these drugs in mice and found a small number could break up the CTC clusters effectively.

The mice treated with these drugs had a significant reduction in metastatic growth compared with untreated mice, while the spread of cancer was nearly eliminated.

'Without our system they could not have even started doing the research,' said Mr Newland.

'This knowledge about CTC clusters is important because more than 90% of people dying from cancer die from metastatic spread of the disease, not the original tumour.

Scientists at Basel hope to start clinical trials in humans with breast cancer this year. If successful, Mr Newland said Angle would try to market the blood testing kit for this specific use as soon as possible.

'The idea is to give cancer patients the Parsortix blood test to see if the clusters are there and break them up. It would be a huge win for Angle if every cancer patient could have this test as a precursor to deciding treatment.'

The Parsortix system has not yet been commercialised and is only being used by scientists and doctors as part of research. However, it is being reviewed by the US drugs regulator and could win approval this year for use in detecting and harvesting for analysis cancer cells in the blood of women with breast cancer.

The hope is that Parsortix will one day be used to routinely monitor cancer patients and help determine if tumours in the body of otherwise healthy people are malignant or benign.

'I'm hoping this work in Basel might do the same for cancer as anti-retroviral drugs did for HIV. It is absolutely amazing news for us and has a huge impact on our revenue potential,'said Mr Newland.

10 January 2019

The dangers of biohacking 'experiments' and how it could harm your health

An article from The Conversation.

THE CONVERSATION

By Rohan Anand, PhD Candidate in Clinical Research, Queen's University Belfast

Biohacking or 'do it yourself' biology has been on the rise in recent years – it now even has various organised conferences. Following a recent *VICE News* documentary about a start-up company called Ascendance Biomedical – who are self-testing drugs – biohacking has had further exposure outside of its circle of devout followers.

Biohacking is an open innovation and social movement that seeks to further enhance the ability of the human body. This includes humans trying to get cyborg-like features, achieve hyper human senses, and also seek out new medicines and cures for disease via the promotion of self-experimentation.

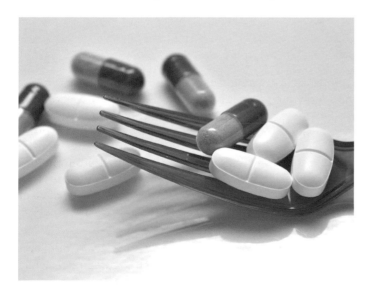

According to their website, Ascendence Biomedical are currently exploring HIV/AIDS and herpes elimination, and 'muscular optimisation'. It sounds futuristic and appealing, but those critical of the approach say a major concern is that the methods of the biohacking community are housed outside of the relevant scientific processes – as governmental, academic, charitable and pharmaceutical institutions that operate with high safety standards for medical research are held to. This means that the biohacking pathway is anything but safe, as it is not regulated.

Why biohack?

Common reasons for biohacking drugs are that there are not enough cures, that drug prices are too high, and that participating in biohacking is taking a stand against the establishment – primarily Big Pharma.

Although modern medicine has progressed rapidly in the last few decades we are still left without cures for many diseases especially chronic, debilitating conditions such as multiple sclerosis or certain cancers. It is natural that anyone suffering from such a disease would be desperate to rid their symptoms and be healthy.

The average cost of getting a drug out of the lab and to patients is US$2.6 billion, and on average it takes around 12 years of research. The process is expensive and slow and it's estimated that less than 1% of candidate drugs get approved.

The research costs of these drugs are also passed on to the patients, meaning they can pay a high price for treatment. And with patent protection, steep prices and years of waiting for cures, it's easy to see why people get frustrated and try to take this process into their own hands.

Why is it dangerous?

In essence, trying to discover drugs through biohacking compromises on quality scientific research. The drugs usually skip key toxicity tests before being administered to patients and in doing so seriously jeopardise the safety of those involved. Without rigorous preclinical testing in the laboratory, it is very difficult to predict how that drug will fully interact with the complexity of the human body.

Gene therapies pose another complexity, they aim to introduce new genetic material into our DNA, essentially rewriting our biological instructions. Edit the wrong part of DNA and you run the risk of seriously interfering with your body such as inducing a tumour. Watching those injecting themselves with unapproved gene therapies is unsettling. And there's also the issue that conclusions drawn from such biohacking 'experiments' are far from evidence-based medicine.

The Government cannot intervene if an individual chooses to self-experiment. And while it's illegal for a company to market something as medicine if it hasn't been approved, chemicals can still be sold as research compounds.

How to carry out medical research

It is vital not to skimp on medical research. Multiple lab experiments are needed to discover the complex mechanisms of drugs and gene therapies to determine if they are safe for humans. Then human testing is best conducted through a series of clinical trials, where each aspect is tightly regulated to ensure scientific integrity and most importantly patients that are safeguarded.

Such trials require an increasingly multidisciplinary team including medics, nurses, methodologists and statisticians to set up and conduct the trial. These trials can minimise bias – for example by using placebo controls. The right number of patients means enough data also allows for statistical validity and legitimate conclusions to be made. Currently this process can be long and expensive, but it produces quality data as to best answer the question of whether a drug or treatment will work.

That said, trials are becoming more efficiently designed and programmes are in place in the UK and US to accelerate drug discovery. Each year the boundaries of medical knowledge are pushed. So things are getting better.

26 September 2018

Seven ways the 'biological century' will transform healthcare

By Mike Moradi, Co-Founder and CEO, Sensulin and David Berry, General Partner, Flagship Pioneering

Among the many disruptive forces that the Fourth Industrial Revolution promises to unleash is a revolution in healthcare. We're entering the century of biology – in which up-to-now crude interventions in our genetic makeup will become far more sophisticated, allowing us to fine-tune our essential biological structure.

These changes will scare some, excite others, and generally raise questions that cannot be answered. Category 5 change, you might say. As well as the philosophical implications, there are practical ones, too, as healthcare worldwide undergoes a total transformation.

Here are our predictions:

1. Death will be optional

The Hayflick limit describes the theoretical restrictions on human life, partially based upon the length of one's telomeres (roughly 125 years of age). To put it another way, cell death is simply a defect in our genes. This defect may be remedied by techniques that are now beginning to make effective medicines for once, intractable diseases.

2. Biotech, like pharma, will be displaced

40 years ago, the founding of Genentech heralded a new era, one overlooked by pharma. Biotech grew to become a more efficient, highly powerful way to develop drugs, while pharma continued down the same path, now operating with a productivity lower than the cost of capital. Two driving forces will again reshape drug development: computational biology and distributed development.

At the intersection of life sciences and supercomputers is computational biology, the use of biological data to frame biomedical problems as computational problems. This has already begun to reshape therapeutic development and is poised to induce a seismic shift in data integration and a new era of disease insight.

In parallel, the next wave will be led by a distributed labs around the world, including China and other regions; medicine will be no longer dominated by the US and Europe. The regulatory environment in the US and Europe makes clinical trials somewhat difficult, and for good reason. However, China has a more open view of regulation in many new areas of medicine, coupled with the world's largest market for most diseases and medical procedures.

3. Our relationship to disease will fundamentally change

Today, our ability to manipulate genes is in its infancy. Nonetheless, dozens of diseases that once were a death sentence are becoming treatable. As our understanding of genes deepens rapidly, less invasive techniques will emerge, starting with native gene control using the body's own biology to cure disease, rather than inserting genes and hoping. Further, the emergence of *in utero* and pre-conception gene sequencing enables the elimination of many congenital birth disorders by sorting rather than editing – sex may even evolve to be solely for recreation, not procreation. Expect a massive *in-vitro* fertilization (IVF) company to emerge in China that will challenge Western ethics.

'Native gene control' could use the body's own biology to fight disease.

4. The future of medicine belongs to smaller, nimbler organisations

According to a recent paper that analysed 121 blockbuster drugs (which made more than $1billion sales per year) between 1980 and 2010, 87.6% were marketed by a major pharmaceutical firm. However, 24.7% were originated by a non-major firm and licensed to big pharma, 37.1% were originated by a non-major firm and acquired or merged into big pharma, and 4.1% were licensed from an academic institution. Thus, 65.9% or two-thirds of all blockbusters originated outside of big pharma. The pharmaceutical industry will continue to exist, as it sheds research jobs and consolidates. However, as capital becomes more available and talent more fluid, smaller organisations will be able to launch new products without big pharma.

5. Previously unimaginable wealth will be created

Let's say that you are the entrepreneur who owns the rights to telomere extension, mentioned above. If you sell this process at an average of $1 per day to three billion people, that amounts to $1.1 trillion a year in revenue. It could also be a completely different innovation, such as the development of human augmentation, instant sleep, instant learning, or something just now having its 'eureka' moment in the cramped quarters of a Tokyo graduate student's lab. We predict that the world's first trillionaire will be a biotech or healthcare entrepreneur. Healthcare accounts for over 20% of the economy in the US, and due to its entrenched bureaucracy, it is a market ripe for disruption. Expect future tycoons to emerge from this chaos.

6. Drugs will get smarter

The joke in venture capital circles is that you can triple your company's valuation by adding cell or gene therapy, immune-oncology or artificial intelligence to your elevator pitch. However, many breakthroughs will emerge from less dramatic advances, including drug delivery technologies such as Sensulin's 'stimulus responsive drug delivery'

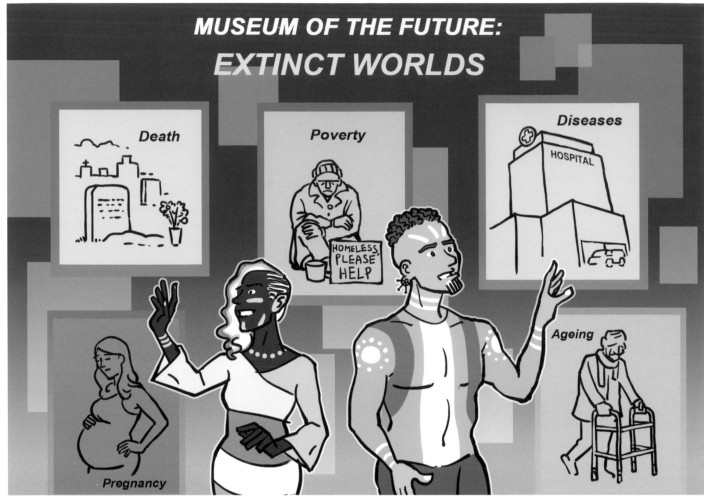

MUSEUM OF THE FUTURE: EXTINCT WORLDS

Death

Poverty

Diseases

HOSPITAL

HOMELESS PLEASE HELP

Pregnancy

Ageing

platform. This technology enables a once-a-day insulin, versus the four to eight injections typically required by patients with type 1 diabetes.

In the year 2000, during the Human Genome Project, the cost of sequencing a human genome was $100 million. Today, the cost is $1,000, or a mere $99 if you only want a partial sequence. Expect similar exponential leaps forward in the coming decades, where all drugs are personalised.

7. Life sciences will permeate most industries.

The tools of biology enable previously unimaginable impact in even some of the most distant industries including, for example, a fascinating company that uses microbes to replace toxic chemicals in the extraction of metals, increasing yields and reducing environmental impact. The company is well capitalised, and their technology is a 'no-brainer' for mining and metals companies. We believe these techniques will ripple across supply chains worldwide, in surprising ways.

The end result?

For the coming biotech wave, the change to our lives will be dramatic. What is certain is that new winners and losers will result. We foresee great leaps in human health, the mitigation of today's chronic diseases, a great number of new jobs created and massive leaps forward in quality of life.

But change will come at a price. Societal conventions will be challenged, increasing inequalities that tear at the very fabric of our societies. Perhaps most poignant, this change

will make us question what it means to be human. What happens in a world where disease is avoidable? Where death is optional?

Preparing for the future

We do not pretend to have all of the answers, but let us ask this crucial question: what must we do in the next decade, to prepare for the coming biotechnology wave?

For starters, scientific literacy is key for our broader population and especially for our leaders. Higher levels of scientific literacy tend to correlate with better governance; thus we urge those in democratic countries to elect leaders who are data-driven in their decision making.

Secondly, we urge an inclusive dialogue between scientists, regulators, elected officials and the general public. Without informed guidance, there will be chaos.

Advances in biotech may not be readily noticeable for some time; though once the proverbial cat is out of the bag, the changes will be too late to influence in a meaningful way. Let us address these challenges while we still can.

1 November 2018

Northern white rhinos could be brought back from brink of extinction 'within three years' using lab-grown embryos

By Josh Gabbatiss

'I do not want to witness northern white rhinos disappear in front of our eyes just because we did not care.'

IVF-ready embryos have been created using sperm from northern white rhinos in an unprecedented development that paves the way for the resurrection of the species.

The sperm was used to fertilise eggs from the closely related southern variety, and the resulting hybrid embryos have been frozen for implantation into surrogate mothers.

Northern white rhinos were left functionally extinct after the last male, Sudan, passed away in March, but their tragic demise has stimulated efforts to develop technologies that can bring them back from the dead.

Using IVF, the researchers hope to see the first northern white calf born for decades within the next three years.

The downward spiral of the rhinos caused by poaching in central Africa has been watched helplessly by scientists and conservationists for years.

As numbers dwindled and the remaining rhinos struggled to reproduce, Dr Thomas Hildebrandt and his team at the Leibniz Institute for Zoo and Wildlife Research collected semen from the last surviving males in the hope it could one day revive the northern whites.

'We came to the point around 2008 that there was no chance to save this subspecies with the techniques we had available at that time,' he said.

These fears were confirmed when tests revealed the only surviving northern whites – a mother and daughter named Najin and Fatu – had serious reproductive problems.

That should have been the end, but rapid advances in reproductive and stem cell science have given Dr Hildebrandt and his international team of experts hope.

Using techniques normally reserved for the creation of champion racehorses, the scientists used some of the preserved sperm to fertilise an egg extracted from a southern white female. The trial served as a test before experimenting with precious eggs taken from northern whites.

The next step is to gain permission from the authorities in Kenya – where Najin and Fatu currently reside – to perform the same procedure using those eggs.

Having created the first-ever test tube rhino embryos, the scientists now see a clear path ahead to creating a new generation of northern whites using surrogate mothers in Africa, Europe and the US.

'We are quite confident with the technology,' said Dr Hildebrandt. 'Our goal is that within three years we have the first northern white rhino calf born.'

'Though there are risks associated with the extraction of eggs from adult rhinos, not least anaesthetising a 1,700kg mammal, the scientists are sure northern white calves produced this way will be perfectly healthy.

'We make foals from the best champions around the world and they still become champions, so the fact the embryo is made in the laboratory doesn't mean it's a fake embryo,' said Professor Cesare Galli, a veterinary medic who led the procedures at Avantea medical laboratory in Italy.

These initial results were published in the journal *Nature Communications*.

In a separate venture, the scientists are aiming to produce new sperm and eggs from stem cells, which they hope to create using samples of skin and other tissues collected from 12 different northern whites.

Though this approach will likely take up to a decade to yield results, the scientists consider it a crucial component in their long-term strategy to create a healthy, genetically diverse rhino population.

'The main goal is to have pure northern white rhinos – it doesn't matter whether it goes through stem cells or it goes through the harvest of oocytes [eggs] from the last living donors,' said Jan Stejskal from Dvur Kralove Zoo, which housed the final northern whites before they were moved to Kenya.

The scientists see it as their responsibility to bring back the species, which once played a crucial role in central African ecosystems, using all the tools at their disposal.

'The northern white rhino didn't fail in evolution, it failed because it was not bulletproof,' said Dr Hildebrandt.

'To me if we have a chance to save them I do not understand why we should not – I do not want to witness northern white rhinos disappear in front of our eyes just because we did not care,' said Mr Stejskal.

4 July 2018

Can genetic engineering save disappearing forests?

An article from The Conversation.

THE CONVERSATION

By Jason A. Delborne, Associate Professor of Science, Policy and Society in the Department of Forestry and Environmental Resources, North Carolina State University

Compared to gene-edited babies in China and ambitious projects to rescue woolly mammoths from extinction, biotech trees might sound pretty tame.

But releasing genetically engineered trees into forests to counter threats to forest health represents a new frontier in biotechnology. Even as the techniques of molecular biology have advanced, humans have not yet released a genetically engineered plant that is intended to spread and persist in an unmanaged environment. Biotech trees – genetically engineered or gene-edited – offer just that possibility.

One thing is clear: the threats facing our forests are many, and the health of these ecosystems is getting worse. A 2012 assessment by the US Forest Service estimated that nearly 7% percent of forests nationwide are in danger of losing at least a quarter of their tree vegetation by 2027. This estimate may not sound too worrisome, but it is 40% higher than the previous estimate made just six years earlier.

In 2018, at the request of several US federal agencies and the US Endowment for Forestry and Communities, the National Academies of Sciences, Engineering, and Medicine formed a committee to 'examine the potential use of biotechnology to mitigate threats to forest tree health'. Experts, including me, a social scientist focused on emerging biotechnologies, were asked to 'identify the ecological, ethical, and social implications of deploying biotechnology in forests, and develop a research agenda to address knowledge gaps'.

Our committee members came from universities, federal agencies and NGOs and represented a range of disciplines: molecular biology, economics, forest ecology, law, tree breeding, ethics, population genetics and sociology. All of these perspectives were important for considering the many aspects and challenges of using biotechnology to improve forest health.

More than 80 million acres are at risk of losing at least 25% of tree vegetation between 2013 and 2027 due to insects and diseases. *Krist et al. (2014), CC BY-SA*

A crisis in US forests

Climate change is just the tip of the iceberg. Forests face higher temperatures and droughts and more pests. As goods and people move around the globe, even more insects and pathogens hitchhike into our forests.

We focused on four case studies to illustrate the breadth of forest threats. The emerald ash borer arrived from Asia and causes severe mortality in five species of ash trees. First detected on US soil in 2002, it had spread to 31 states as of May 2018. Whitebark pine, a keystone and foundational species in high elevations of the US and Canada, is under attack by the native mountain pine beetle and an introduced fungus. Over half of whitebark pine in the northern US and Canada have died.

Poplar trees are important to riparian ecosystems as well as for the forest products industry. A native fungal pathogen, *Septoria musiva*, has begun moving west, attacking natural populations of black cottonwood in Pacific Northwest forests and intensively cultivated hybrid poplar in Ontario. And the infamous chestnut blight, a fungus accidentally introduced from Asia to North America in the late 1800s, wiped out billions of American chestnut trees.

Can biotech come to the rescue? Should it?

It's complicated

Although there are many potential applications of biotechnology in forests, such as genetically engineering insect pests to suppress their populations, we focused specifically on biotech trees that could resist pests and pathogens. Through genetic engineering, for example, researchers could insert genes, from a similar or unrelated species, that help a tree tolerate or fight an insect or fungus.

It's tempting to assume that the buzz and enthusiasm for gene editing will guarantee quick, easy and cheap solutions to these problems. But making a biotech tree will not be easy. Trees are large and long-lived, which means that research to test the durability and stability of an introduced trait will be expensive and take decades or longer. We also don't know

nearly as much about the complex and enormous genomes of trees, compared to lab favourites such as fruit flies and the mustard plant, *Arabidopsis thaliana*.

In addition, because trees need to survive over time and adapt to changing environments, it is essential to preserve and incorporate their existing genetic diversity into any 'new' tree. Through evolutionary processes, tree populations already have many important adaptations to varied threats, and losing those could be disastrous. So even the fanciest biotech tree will ultimately depend on a thoughtful and deliberate breeding programme to ensure long-term survival. For these reasons, the National Academies of Sciences, Engineering, and Medicine committee recommends increasing investment not just in biotechnology research, but also in tree breeding, forest ecology and population genetics.

Oversight challenges

The committee found that the US Coordinated Framework for Regulation of Biotechnology, which distributes federal oversight of biotechnology products among agencies such as EPA, USDA and FDA, is not fully prepared to consider the introduction of a biotech tree to improve forest health.

Most obviously, regulators have always required containment of pollen and seeds during biotech field trials to avoid the escape of genetic material. For example, the biotech chestnut was not allowed to flower to ensure that transgenic pollen wouldn't blow across the landscape during field trials. But if biotech trees are intended to spread their new traits, via seeds and pollen, to introduce pest resistance across landscapes, then studies of wild reproduction will be necessary. These are not currently allowed until a biotech tree is fully deregulated.

Another shortcoming of the current framework is that some biotech trees may not require any special review at all. The USDA, for example, was asked to consider a loblolly pine that was genetically engineered for greater wood density. But because USDA's regulatory authority stems from its oversight of plant pest risks, it decided that it did not have any regulatory authority over that biotech tree. Similar questions remain regarding organisms whose genes are edited using new tools such as CRISPR.

The committee noted that US regulations fail to promote a comprehensive consideration of forest health. Although the National Environmental Policy Act sometimes helps, some risks and many potential benefits are unlikely to be evaluated. This is the case for biotech trees as well as other tools to counter pests and pathogens, such as tree breeding, pesticides and site management practices.

How do you measure the value of a forest?

The National Academies of Sciences, Engineering, and Medicine report suggests an 'ecosystem services' framework for considering the various ways that trees and forests provide value to humans. These range from extraction of forest products to the use of forests for recreation to the ecological services a forest provides – water purification, species protection and carbon storage.

The committee also acknowledged that some ways of valuing the forest do not fit into the ecosystem services framework. For example, if forests are seen by some to have 'intrinsic value', then they have value in and of themselves, apart from the way humans value them and perhaps implying a kind of moral obligation to protect and respect them. Issues of 'wildness' and 'naturalness' also surface.

Wild nature?

Paradoxically, a biotech tree could increase and decrease wildness. If wildness depends upon a lack of human intervention, then a biotech tree will reduce the wildness of a forest. But perhaps so would a conventionally bred, hybrid tree that was deliberately introduced into an ecosystem.

Which would reduce wildness more – the introduction of a biotech tree or the eradication of an important tree species? There are no right or wrong answers to these questions, but they remind us of the complexity of decisions to use technology to enhance 'nature'.

This complexity points to a key recommendation of the National Academies of Sciences, Engineering, and Medicine report: dialogue among experts, stakeholders and communities about how to value forests, assess the risks and potential benefits of biotech, and understand complex public responses to any potential interventions, including those involving biotechnology. These processes need to be respectful, deliberative, transparent and inclusive.

Such processes, such as a 2018 stakeholder workshop on the biotech chestnut, will not erase conflict or even guarantee consensus, but they have the potential to create insight and understanding that can feed into democratic decisions that are informed by expert knowledge and public values.

18 January 2019

You won't believe these three unexpected discoveries – and neither did the scientists who made them

Researchers at one of the the UK's leading genetics centres tell us about their serendipitous findings.

By Gaia Vince

Science, mostly, progresses iteratively. But every now and then, a discovery will be made – often incidental to the main aim of the research – that is entirely unexpected. Such serendipitous findings enable us to leapfrog our usual incremental advances. They can even disrupt a whole field of research.

In 2003, the Human Genome Project, the 13-year-long international effort to fully sequence human DNA and identify all our genes, was completed. The Wellcome Sanger Institute, near Cambridge, England, was the only British organisation involved, completing the sequence of one-third of the genome.

While many of the great proclamations made at the launch of the project have yet to be realised, there is no doubt that sequencing the human genome was a technological gamechanger for science in the way that, say, the invention of the printing press and microscopes were in previous centuries. And, just as for other disrupting technologies, genome sequencing has led to some wholly unexpected findings.

1. Chromosome shattering can cause cancer

Cancer is a genetic disease. Through replication errors and mutations in DNA, healthy cells form tumours that can kill us. Today, researchers are sequencing the genomes of many tumours and through this transforming our understanding of what exactly cancer is.

Peter Campbell, who heads Sanger's cancer programme, has made a few unexpected discoveries in his work with kidney tumours, but the most remarkable was an entirely new cancer-triggering mechanism.

He and his colleagues found that a chromosome can explode for unknown reasons, shattering into hundreds of pieces.

It was such a surprising finding that Campbell assumed it was because of a problem with the data. 'Almost all of these things turn out to be rubbish in the data – someone's mucked something up somewhere along the line. So usually it starts with us saying: we need to figure out what's gone wrong,' he says. However, no obvious error could be found.

'The advantage of working with genetic data is it's black and white, it behaves in a digital way – unlike, say, cells, which can look different on different days.'

Once he'd confirmed the discovery, Campbell allowed himself to get excited. Science, he notes, is usually a very long-term project: 'You can be banging your head against problems that you can't solve for weeks on end… then unexpected findings drop out all of a sudden.'

Campbell prepared himself for a backlash from the field. In the end, though, his research paper was received very positively, and others have since confirmed his findings, capturing the changes occurring in cells grown in the laboratory. The mechanism is likely to trigger many different cancers, researchers believe.

So has the finding shifted people's thinking? 'Yeah, I think so,' says Peter. 'The best papers do that, but they come along not so often in one's own career. We were lucky that we had very early access to the modern genetic technologies that allowed us to spot these patterns before others did.'

2. The Y chromosome is not useless after all

Haematologist George Vassiliou and his team were looking for new targets for drugs to treat leukaemia. One such target was a cancer-suppressing gene called UTX, which sits on the X chromosome.

As part of their investigations, they were using mice with faulty versions of the UTX gene to see when they got cancer. During these experiments, Vassiliou also tweaked a similar nonfunctional gene on the Y chromosome called UTY.

Aside from the gene that determines sex, Y chromosome genes were largely thought to be non-functional leftovers. In fact, the mole vole has evolved to lose its Y chromosome altogether.

Researchers have suggested that the same could happen in humans eventually. That was until Vassiliou's team discovered that the UTY gene is functional in humans. What's more, they found that it plays a significant role in suppressing cancer. 'It was very exciting,' he says. 'At first, we were so happy to find another potential target for leukaemia treatment.'

But then there was a moment of realisation: 'The Y chromosome actually does something! It is not useless after all.'

3. Bacterial genes we thought might not do anything actually affect how well vaccines work

Jukka Corander, a biostatistician at the Sanger Institute and University of Oslo, and Nicholas Croucher, a bacterial geneticist at Imperial College London, were exploring the genomics of bacterial infections. They used computer simulations of multiple strains to figure out what causes bacterial populations to change in unexpected ways after vaccination against them.

They were comparing the genomes of *Streptococcus pneumoniae*, a bacterium that causes severe illnesses such as pneumonia, sepsis and meningitis. This involved collections of bacteria from four different human populations around the world, three of which had been vaccinated against the bacterium.

All *Streptococcus pneumoniae* strains have around 2,000 genes. Three-quarters of these genes are very similar across strains. The remaining 'accessory genes' vary considerably between them. Bacterial strains can swap any gene through a process called horizontal gene transfer.

The modelling showed that the levels of accessory genes were very similar in the bacterial population before and after vaccination, even though the types of strain present had changed dramatically.

In other words, the strains that emerged after vaccination had similar sets of accessory genes to the strains eliminated by vaccination. So – far from being useless – accessory genes appear to play a role in how the bacterial population responds to a vaccine.

Importantly, some of the accessory genes that returned to their previous levels are involved in resistance to antibiotics.

'We had just published a paper that agreed with the 'neutral model' of the accessory genes,' Corander says, 'so I was utterly astonished.'

This work provides the foundation for further research that will help predict which strains will spread most rapidly after a change to how we treat bacterial diseases.

'Without having access to high-definition genomes, we would never have seen this,' Corander says. "We wouldn't even have known that so much variation exists in the genomes, let alone this important role of these rare accessory genes.'

11 September 2018

Wellcome, the publisher of Mosaic, founded the Wellcome Sanger Institute in 1993 and has funded it ever since. The Sanger Institute celebrates its 25th anniversary in October 2018.
Nicholas Croucher holds a Sir Henry Dale Fellowship, which is funded by Wellcome and the Royal Society. Peter Campbell receives funding from Wellcome through a Senior Research Fellowship in Clinical Science.

The most pressing issues in bioethics

By The Medical Futurist

Who owns medical and genetic data? How to regulate gene editing? Where is the boundary of enhancing physical or cognitive human capabilities? What to do with biological differences widening the gap of the haves and have-nots? Could we define where the boundary is to augment life? Will we sue robots or algorithms for medical malpractice? With the constant advancement of technology, unprecedented moral, ethical and legal concerns are surfacing.

Channelling them into substantial debates will get us closer to their fair solution step by step. Here, we collected the most pressing issues in bioethics.

Bioethicists of the world, unite!

In November 2018, a scientist in China claimed to have edited a gene in two human embryos and implanted them in their mother's womb, resulting in the birth of genetically altered twin girls. The case caused international outrage in scientific circles, and a couple of months later, many scientists called for a global moratorium on gene editing embryos. However, that's just one story from the ethically challenging situations on the edges of medical scientific innovation.

The billionaires of Silicon Valley, as well as other prime examples of the 'haves', are pouring money into research of longevity and aging, which might result in the even-more widening gap between the life expectancies of the affluent and the masses. Dr David Himmelstein, the co-founder of the Physicians for a National Health Program and a lecturer in medicine at Harvard Medical School, told Healthline that the gap between the wealthiest and poorest Americans is already about ten years for women and 15 for men – and we could expect an exponential increase there.

Besides, we haven't even touched upon issues around private and sensitive medical data, not to say genetic data. Who's the owner, the distributor or the user of that data? What is it worth and who uses it for what purposes? Or what about connected medical devices used for healthy people? Could the use of an exoskeleton be allowed in a warehouse to enhance capabilities? What about brain implants or digital tattoos? Where should or might the augmentation of human bodies or cyborgisation stop?

We're no longer in the realm of science fiction. These issues are real, well and alive. While *The Medical Futurist* takes an overall optimistic approach to technology and emphasizes its benefits, we have to deal with the dark side – in order to take steps to counter them as soon as possible. Physicians, patients, regulators, and all other stakeholders must prepare for the coming waves of change. To do that, we

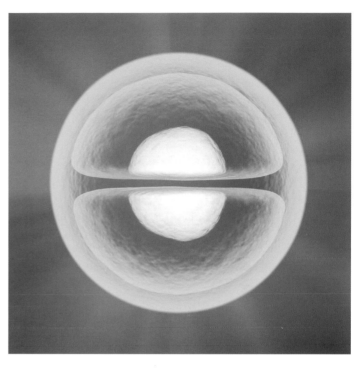

need to start talking openly about the dangers we face. To spark discourse and urge bioethicists to speak a lot more and a lot louder about the most pressing issues in bioethics, here's the ultimate list about the ones we consider the most relevant ones.

1) Medical and genetic data privacy

The most important bioethical issue of our times is how to treat data, more specifically how to treat private and sensitive medical and genetic data. How should we secure, share or trade with sensitive data? Should sensitive fitness and medical data be shared with insurance companies? What if you ate red meat and your insurance company immediately raised your insurance rates because you're not eating healthy enough?

What to do with genetic and genomic data? If you already purchased a direct-to-consumer genetic test or any other one, what can companies do with your data? Can you delete it or force the companies to make your genetic information disappear? Could it be allowed for companies, research conglomerates or pharma enterprises to purchase or sell such data, and if anyone does that, how to set the price for it? What is the price for a single person's sequenced genome? Should there be a price?

These are not far away and remote concerns. For example, in July 2018, GlaxoSmithKline decided to invest $300 million in 23andMe and forge an exclusive drug development deal with the Silicon Valley consumer genetics company to research and develop innovative new medicines and potential cures, using human genetics as the basis for discovery. Ancestry, which maintains a more than five-million-person consumer database of genetic information, once partnered with Google's stealthy life-extension spinoff Calico to study aging. Caitlin Curtis, a research fellow at the University of Queensland, estimates 23andMe made around $130 million from selling access to about a million genotypes, before the GSK deal, implying an average price of around $130. That means if you purchased 23andMe's genetic test for $100 – 150, your genetic information could have been bought for another $130 on average price. The question is whether we are okay with that…

2) Cyber attacks against medical devices and systems

Do you remember the WannaCry scandal, the global cyber attack that infected 300,000 computers in 150 countries using hacking tools? It also crippled the National Health Service (NHS) in the United Kingdom. UK hospitals were shut down and had to turn away non-emergency patients after ransomware ransacked its networks. That was the complete and utter failure of the health IT infrastructure.

Since that attack, not only hospitals doubled down on cybersecurity, but Microsoft also started to take cybersecurity in healthcare as seriously as never before. The ransomware exploited a vulnerability that Microsoft had created a patch for two months earlier, but many organisations – including hospitals – had not appropriately updated their systems before the attack. But is okaying the updates enough? And did huge medical facilities or tech companies dealing with vulnerable data even do

the homework ever since? It doesn't seem so, although they should! Recently, it turned out that California-based Meditab, a health tech company was leaking thousands of doctor's notes, medical recordsand prescriptions daily after a security lapse left a fax server without a password.

The situation is not rosy regarding the security of medical devices either. In 2011, a researcher from the McAfee tech company demonstrated at a conference in Miami how insulin pumps might be hacked to deliver fatal doses to diabetes patients. According to the latest news, Homeland Security has issued a warning for a set of critical-rated vulnerabilities in Medtronic defibrillators that put the devices at risk of manipulation. The question is – what can we do to protect wearable devices that are connected to our physiological system from being hacked and controlled from a distance? Companies developing such technologies should make sure they are safe and users should be as vigilant as possible when using them.

3) What to do with biohackers?

With the rise of the maker movement, the availability of know-how, raw materials and an active community, the appearance of 'garage solutions' in medicine have multiplied. However, they cannot be considered as unambiguously positive or negative. Some people have been working out long-term solutions for serious medical conditions outside of the traditional 'ivory tower of medical knowledge' as they considered regulation to be too slow compared to innovation.

Our favourite example is the #wearenotwaiting Twitter-movement for patients suffering from diabetes. The initiator of the community, Dana Lewis, and her husband built a so-called artificial pancreas at home and started to spread the blueprints and know-how on Twitter to other diabetes patients – without waiting for the approval of the FDA or any other agency.

Why? Because it works and patients needed it. Dana had been using the device for almost two years by the time the US Food and Drug Administration finally approved it.

On the other hand, experimenting at home with unapproved or not well-tested technologies is dangerous. Does it mean that patients who will be able to scan themselves, 3D print medications at home or even do genetic engineering should also be allowed to do all these? How should we go about biohackers? Where should the line be drawn between supporting innovation and refusing reckless experimentation? Do you think that it's just another piece of far-fetched, non-existent gobbledygook? Have you heard about the biohacker who tried to do CRISPR therapy on himself at his own home? Josiah Zayner injected his arm with DNA encoding for CRISPR that could theoretically enhance his muscles – in between taking swigs of Scotch at a live-streamed event during an October conference. Now, he believes that wasn't a good idea.

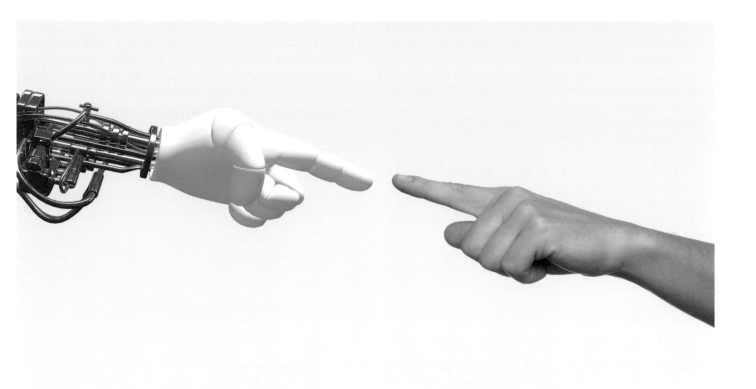

4) What if healthy people turned to technology?

As technological innovations in the field of medicine and healthcare multiply day by day, it will be more and more usual to augment our bodies with the help of machines. It makes us faster, stronger or more sensitive to the environment. This means that the boundaries of 'human-ness' are stretched. How far can we and should we go?

At first, we might experiment with exoskeletons that let warehouse workers lift heavy boxes or allow doctors to stand through operations for more than 10 hours. Those are easily removable mechanic extensions, but what about digital tattoos or other implants? What if someone wants to have a brain implant which lets him experience sensations better? We surely know that could go wrong – just think of the horrible episode of *Black Mirror*, Black Museum.

However, digital tattoos could also be removed upon request, we assume, but what if people go one step even further in cyborgisation and request irreversible changes in their bodies? What if people start asking their doctors to replace their healthy limbs for robotic ones because it would let them run faster? What if they start asking for indelible brain chips to get smarter? Currently, you can get a new nose or larger breasts, what would prevent you from getting new muscles or brain implants?

5) Biological differences based on inequalities in wealth

Both staying healthy and healing from a condition cost enormous sums of money (well, for the average person, not the Bugatti-buying upper 0.1%). Americans are even worse off than other developed nations as the US spent 17.8% of its GDP on health care in 2016. Meanwhile, the average

spending of 11 high-income countries assessed in a report published in the *Journal of the American Medical Association* – Canada, Germany, Australia, the UK, Japan, Sweden, France, the Netherlands, Switzerland, Denmark, and the US – was only 11.5%.

Moreover, the average cost of hospital stays for cancer patients in 2015 was $31,390, according to US government figures – about half that year's median household income. In addition, medical expenses are the leading cause of bankruptcy in the US, according to a study that indicated about 62% of personal bankruptcies in 2007 were reportedly due to medical bills, even though most of those people had insurance – up from about 46% in 2001.

It is even widely reported that the differences in the financial background of people cause visible biological differences. As mentioned above, the gap between the wealthiest and poorest Americans is about ten years for women and 15 for men, so there is a big spread with more affluent people living much longer than the poorer masses. And some are even expecting the widening of the chasm. With the appearance of direct-to-consumer genetics, (for some) affordable whole-genome sequencing, and later on technologies which can truly augment human capabilities – exoskeletons, implants, digital tattoos, artificial limbs and so on – people with the appropriate means will live longer and healthier lives. How can we mitigate the differences? How can we make innovations more accessible to all kinds of communities? At the same time, how do we prepare society for a time when financial differences lead to biological ones?

6) What if we live beyond 130 years?

Since 1840, life expectancy at birth has risen about three months per year. Thus, every year a newborn lives three months longer than those born the previous year. Sweden, which keeps excellent demographic records, documents female life expectancy at 45 years of age in 1840 and 83 today. Experts even believe that with recent breakthroughs in science and medicine coupled with lifestyle changes, this number could reach far beyond 100 years. Tons of ethical and philosophical questions appear with that possibility.

What would longevity bring for the individual and for society? Does a longer life also go hand in hand with a physically and cognitively stable older age? Do we even want to live longer if we cannot keep our bodies fit for the task? And what about our societies? How would governments, institutions, communities and even our ideas about life itself cope with the changes? How could we extend our lifespan beyond 100 years of age if the effects of an aging population already strain our societies? If younger generations cannot sustain the social system to provide care for their elders as they are growing older and older, significant structural changes will be necessary. Are we ready for those?

7) The horrors of bioterrorism

The sensitivity of medical and genetic data is mostly due to fears that they could end up in the wrong hands – and for the time being, nothing can be done to counter a potential attack. Although wrongdoers have to possess very sophisticated skills, so the risks are very low, there are some experts who worry that precision bioterrorism could appear alongside precision medicine and targeted treatments. That would mean that according to genetic markers or any other biological markers, attackers could choose a target population and tailor their biological attack according to their genetic makeup or medical data. As you cannot change your genome as you do with your passwords or credit cards, anyone can be completely defenseless against such an attack. Some FBI agents reported they worry about healthcare data generated for precision medicine leaving the US vulnerable to such scenarios.

Alongside with that, hacking of medical devices, complete infrastructures, and systems, manipulation of implants, digital tattoos or robots might leave us utterly assailable. As in the far future, robots on the nanoscale could live in our bloodstream or on our eyeballs, some people are also afraid that by using such tiny devices, total surveillance would

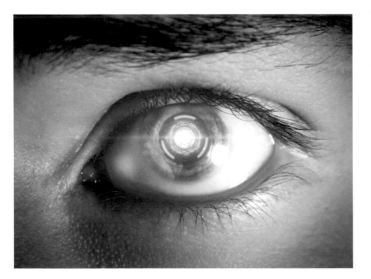

become feasible – since nothing can remain hidden when there is a robot swimming through your bodily fluids. Let's do everything in our power to counter these risks and come up with defensive measures as soon as possible.

8) Sexuality becoming technological

Long-distance kisses, hugs and caresses. Virtual reality porn stars. Sex robots threatening the world's oldest profession. Technosexuals living with life-sized dolls. At the dawn of a new sexual revolution, it's time to face where technology may take the most intimate area of our lives, where will that take humanity, where are our boundaries and whether we want any of it.

According to the *Future of Sex* report written by Jenna Owsianik and Ross Dawson, haptic body suits or social networking sites enhanced with sensual touch interfaces might soon enable fully physical long-distance sex between lovers or hook-up partners thousands of kilometres away. Not to mention the possibility of intimate video chats combining remote sex devices and holographic partners. The authors even estimate that by 2028, that means in ten years (!), over a quarter of young people will have had a long-distance sexual experience. By 2025, 3D-printed body parts could add more realism to the over-the-line sex game.

The other line of innovation includes robots resembling human sex partners. Some futurists even speculated that by 2050, human-robot sex will be more common than human-human sex. But can you imagine having sex with a remote program or a robot?

And what if something goes wrong? Do you see such, right now seemingly impossible future news headlines as 'Woman Sues Robotics Company for Breaking Her Leg During Sex With Robot'?

These questions cannot be answered by 90-minute-long keynotes and a Q&A session afterwards. All kinds of stakeholders need to be brought together by bioethicists to see a vivid palette of opinions, after which, the potential ethical, moral or even legal rules can be set up concerning each of the issues. That doesn't mean only formal discussions – let's discuss these bioethical issues at home, at the workplace and on public forums. This way, we can prepare to exploit the advantages technology offers, while keeping the potential dangers at bay.

And we shall never forget: *'Primum non nocere'*.

26 March 2019

Scientists disagree about the ethics and governance of human germline editing

By Francoise Baylis and Marcy Darnovsky

Despite the appearance of agreement, scientists are not of the same mind about the ethics and governance of human germline editing. A careful review of public comments and published commentaries in top-tier science journals reveals marked differences in perspective. These divergences have significant implications for research practice and policy concerning heritable human genome editing.

The current chapter in the debate about the societal and political implications of human germline editing took off nearly four years ago, in response to a laboratory experiment in which researchers in China used CRISPR technology on nonviable human embryos. In March 2015, an article titled *Don't Edit the Human Germline,* coauthored by scientists and others working on somatic cell genome editing and associated with the Alliance for Regenerative Medicine, appeared in the comment section of *Nature*. A week later, *Science* published a 'Perspective' commentary coauthored by another group, most of them scientists convened by CRISPR codiscoverer Jennifer Doudna, under the title 'A prudent path forward for genomic engineering and germline genetic modification'.

The first article described the tenuous therapeutic benefit, and the likely serious risks (including risks to future generations), of germline editing. The authors concluded that this technology was 'dangerous and ethically unacceptable', in part because 'permitting even unambiguously therapeutic interventions could start us down a path towards non-therapeutic genetic enhancement'. They further suggested that 'a voluntary moratorium in the scientific community could be an effective way to discourage human germline modification".

In contrast, the second article stressed 'the promise of curing genetic disease'. While it concluded that 'there was an urgent need for open discussion of the merits and risks of human genome modification by a broad cohort of scientists, clinicians, social scientists, the general public, and relevant public entities and interest groups', it asserted that the goal of such discussion would be to 'enable pathways to responsible uses of this technology, if any, to be identified'. The authors recognised that it would be premature to move forward with human germline genome editing, but studiously avoided any use of the term 'moratorium'.

The lines drawn by those articles persist to this day. In fact, they are prominently on display in a recent set of responses to the self-reported experimentation by He Jiankui in attempting to create gene-edited babies. A *Nature* editorial published soon after He's claims went public notes that a system to regulate powerful new genome editing tools 'should not start with the assumption that future germline editing is a foregone conclusion'. That, it says, 'is a question for society, not scientists, and one that demands the input of different stakeholders from across the world'.

Moreover, it advises that 'researchers and physicians must ask permission rather than beg for forgiveness'.

Meanwhile, an editorial in *Science*, signed by the presidents of three national scientific academies (two in the US and one in China), assumes the very conclusion that others warn about. These authors insist on the need to develop 'criteria and standards' for 'genome editing in human embryos for reproductive purposes', and pointedly note that once these criteria and standards are in place, germline editing for human reproduction will be 'deemed permissible'. Abandoning the commitment to 'broad *societal* consensus' included in an influential statement issued by the organizing committee of the first International Summit on Human Gene Editing in 2015, they call for 'broad *scientific* consensus'.

With this move, they arrogate to those scientists who agree with them the authority to decide whether to proceed with human germline genome editing – no matter that this technology could reshape societies with new divides between genetic haves and have-nots. They make clear that they are committed to voluntary self-regulation, not enforceable public policy – notwithstanding the profound societal shifts that this powerful biotechnology could introduce. Indeed, they write of *'nascent* efforts' to 'confront

Key facts

- There are many different types and applications of biotechnology, including; medical, agricultural, industrial, food, energy production (biofuels) and pharmaceutical development. (pages 1–3))

- Genetically modified organisms (GMOs), including microbes, cells, plants and animals, have long been used in scientific and medical research as a way to understand processes in biology as well as the mechanisms of diseases. (page 4)

- Since the late 2000s, scientists began to develop techniques known as 'genome (or gene) editing'. Genome editing allows scientists to make changes to a specific 'target' site in the genome. One of the techniques that has generated the most excitement due to its efficiency and ease of use, is called 'CRISPR'. CRISPR stands for 'clustered regularly interspaced short palindromic repeats'. (page 5)

- Currently, germline genetic modification is illegal in many European countries and in Canada, and federal funding in the United States cannot be used for such work. As of January 2017, researchers in the UK, Sweden and China have received approval to perform gene editing in human embryos for research purposes only (in addition, existing laws or guidelines in these countries only allow research on embryos up to 14 days after fertilisation). (page 6)

- In November 2018, news reports emerged that the first children whose genomes were edited with CRISPR during their embryonic stage, a pair of twins, have been born in China. While the claims have still not been independently validated or published in peer-reviewed journals, the claims have drawn significant controversy. (page 6)

- Significant medical breakthroughs are happening via synthetic biology. The antimalarial treatment artimisinin can now be produced by yeast, avoiding the need to isolate it from Chinese sweet wormwood plant. This helps to stabilise global prices. (page 7)

- In 2016, a new immune cell engineering treatment resulted in a 50% complete remission rate in terminally ill blood cancer patients, with a 36% remission rate achieved in a 2017 trial. (page 7)

- In 2015, the synthetic biology component market (DNA parts) was worth $US5.5 billion – by 2020, it will approach $US40 billion. (page 8)

- The British public largely support using gene editing to prevent people from passing on hereditary genetic disorders. Three quarters (76%) of the adult population think it should be allowed, with just one in eleven (9%) saying it should be not be allowed. One in six (14%) don't know. (page 9)

- Brits are much more reluctant to support gene editing to boost either the brains or beauty of their future children. Seven in ten (71%) oppose editing genes for intelligence, and three–quarters (76%) are against doing it for appearance. (page 9)

- Around one in six (15%) 18-to-24-year-olds support editing genes affecting intelligence and one in eight (13%)for appearance. (page 10)

- 83% of Brits would consider gene editing if they were carrying genes for a disorder that could be passed on to their offspring. (page 10)

- The 100,000 Genomes Project began in 2012 with the aim to sequence 100,000 whole genomes from around 70,000 participants with rare disease, their families and people with some cancers The decision was backed by robust government support – both political and financial – which included over £300 million of investment. (page 11)

- In a recent survey it was found that 60% of vegans would be happy to eat lab-grown meat. However, the rest of the public are not quite so enthusiastic. Only 18% of UK respondents said they would be up for eating it. (page 19)

- The average cost of getting a drug out of the lab and to patients is US$2.6 billion, and on average it takes around 12 years of research. The process is expensive and slow and it's estimated that less than 1% of candidate drugs get approved. (page 21)

- In the year 2000, during the Human Genome Project, the cost of sequencing a human genome was $100 million. Today, the cost is $1,000, or a mere $99 if you only want a partial sequence. (page 24)

- Medical expenses are the leading cause of bankruptcy in the US, according to a study that indicated about 62% of personal bankruptcies in 2007 were reportedly due to medical bills, even though most of those people had insurance – up from about 46% in 2001. (page 33)

- Since 1840, life expectancy at birth has risen about three months per year. Thus, every year a newborn lives three months longer than those born the previous year. Sweden, which keeps excellent demographic records, documents female life expectancy at 45 years of age in 1840 and 83 today. (page 33)

- The most widely used genetically modified animals are laboratory animals, such as the fruitfly (*Drosophila*) and mice. (page 38)

Bioethics

Bioethics are the considerations of the issues surrounding medical and biological research.

Biohacking

Biohacking is a growing practice also known as DIY biology. Individuals or groups experiment on themselves outside a traditional medical research environment.

Biotechnology

Biotechnology is the use of natural organisms and biological processes to change or manufacture products for human use. Biotechnology is widely used in modern society: for example, in agriculture, pharmaceuticals, the manufacturing industry, food production and forensics.

CRISPR

Pronounced 'crisper', CRISPR is short for Clustered Regularly Interspaced Short Palindromic Repeats and is a new technique for genetically modifying animals (a new way of modifying DNA which can reduce the need for breeding). In 2012, CRISPR was discovered in the primitive immune system of bacteria. It allowed researchers to create a break in the DNA helix at very specific places, where they can then introduce mutations as the break is repaired, making it a much more precise technique. The precision offered by this technique offers an additional level of refinement for researchers – if they can produce better models for diseases then the results will be even more reliable. The technique is still in its infancy, but the potential for CRISPR is massive: it is cheaper, simpler, more reliable and requires fewer animals.

DNA

DNA (deoxyribonucleic acid) is the genetic coding which is present in every cell of living organisms. DNA is found in the nucleus of each cell and determines the characteristics for that organism.

Genes

A gene is an instruction and each of our cells contains tens of thousands of these instructions. In humans, these instructions work together to determine everything from our eye colour to our risk of heart disease. The reason we all have slightly different characteristics is that before we are born our parents' genes get shuffled about at random. The same principles apply to other animals and plants.

Gene editing

A type of gentic engineering in which a selected DNA sequence in a living cell is altered through deletion, insertion or replacement of sections of DNA.

Genetic modification

May also be called modern biotechnology, gene technology, recombinant DNA technology or genetic engineering. Scientists are able to modify genes in order to produce different characteristics in an organism than would have been produced naturally. GM techniques allow specific genes to be transferred from one organism to another, including between non-related species. This technology might be used, for example, to produce plants which are more resistant to pesticides, which have a higher nutritional value or which produce a greater crop yield. Those in favour of GM say that this could bring real benefits to food producers and consumers. Those against GM feel it is risky as scientists do not have the knowledge to 'play God' with the food we eat.

GM food

GM (or genetically modified) foods are products which have undergone a process of genetic selection to produce a desired characteristic. For example, scientists may transfer the gene for disease resistance from one organism into a genetically unrelated crop, which would result in an improved yield from the modified crop.

In vitro meat

Animal flesh cells that have been grown in a laboratory and which have never been part of a living animal. Scientists are developing in vitro meat as a solution to the health and environmental problems associated with natural animal farming.

Nano-technology

The science of manipulating atoms and/or molecules to create materials and devices. This is all done on an extremely small scale – a nanometre is one billionth of a metre. Examples of nano-technology include micro-electronics (think tiny computer chips) and also nano-machines (incredibly small machines such as gears, switches, etc, or even nano-robots).

Selective breeding

Human beings have been modifying the genes of biological organisms for centuries through selective breeding: choosing individual plants and animals with particular traits, like fast growth rates or good seed production, and breeding them with others to produce the most desirable combination of characteristics. However, unlike genetic modification, this can happen only within closely-related species.

Synthetic biology

Synthetic biology (synbio) is an extreme version of genetic engineering. Instead of swapping genes from one species to another (as in conventional genetic engineering), synthetic biologists employ a number of new genetic engineering techniques, such as using synthetic (human-made) DNA to create entirely new forms of life or to 'reprogramme' existing organisms to produce chemicals that they would not produce naturally.

Assignments

Brainstorming

- In small groups, discuss what you know about biotechnology. Consider the following questions:

  - What is biotechnology?

  - What is bioethics?

- In groups, create two spider diagrams on a large sheet of paper (A2 size is ideal). One should include all the reasons you can think of to support the idea of using biotechnology to develop synthetic food, or 'Frankenfoods', such as meat and milk, whilst the other should present the arguments against it. Which diagram is more convincing? Which side of the debate do most people agree with? Feedback to your class and discuss the points on your diagrams.

- As a class, list as many big scientific discoveries and breakthroughs as possible that you are aware of, historical or recent. How many did you come up with? If you came up with a list of ten or more, organise a poll and ask everyone to vote for the top three breakthroughs they think have been the most important, or had the biggest impact on people's lives.

Research

- Create a questionnaire that will explore opinions about gene editing. You should aim to find out:

  - How do people view gene editing – positively or negatively? Why do they feel that way? List the reasons the respondents give in two columns – FOR and AGAINST.

  - Discuss the results in small groups.

  Analyse your results and write a report that includes at least two graphs/visual representations of your data.

- Read through the article on page 21 about biohacking. Do some further reading on the topic and choose a story that peaks your interest and conduct further research. Make notes and present them to your class in a quick two-minute presentation. Look at the 'Useful weblinks' at the front of this book, or visit Issues Online for help with your research.

- Read the article on page 17, 'Could an Australian bee solve the world's plastic crisis?' Conduct some research looking for similar stories where natural plant and animal by products, and 'biomimicry', have been used to solve a problem. Share the examples you have found with the rest of the class.

Design

- Imagine that you work for a company who have created a synthetic meat that looks and tastes like the real thing – it could be chicken, sausages, prawns or fish fingers, for example. Think of a name and a logo for your new product, then design a poster and some sample pages from your company website. You could even suggest some recipes for people to try. Get creative! (Work in groups or pairs.)

Oral

- As a class, or at home, watch the film *Jurassic Park*. If you could persuade a genetic scientist to bring an animal or species back from extinction, what would you choose, and why?

- 'Genetic modification should be illegal, no matter what it is used for.' Stage a class-debate in which half of you argue in favour of the statement and half of you argue against it.

- Biotechnology is an incredibly fast-paced field. Read the article on page 23, 'Seven ways the biological century will transform healthcare' and with a partner or in small groups, think about a biotech breakthrough or innovation you would like to see happen in the next five years.

Reading/writing

- Write a blog post from the point of view of a scientist who is developing synthetic milk. Write about the things you hope to achieve with this development.

- Watch the film *The Island* (2005) and write a review discussing how the director addresses the theme of biotechnology.

- Read the article on page 38 'Is it ethical to genetically modify farm animals for agriculture?' and write a summary for your school newspaper, adding in your own opinions if you feel necessary.

- Imagine you are a science correspondent for an international news agency. Write a short report on a ground-breaking scientific discovery, real or fictitious. What headline would you most like to write?

Acknowledgements

The publisher is grateful for permission to reproduce the material in this book. While every care has been taken to trace and acknowledge copyright, the publisher tenders its apology for any accidental infringement or where copyright has proved untraceable. The publisher would be pleased to come to a suitable arrangement in any such case with the rightful owner.

Images

All images courtesy of iStock except pages 2, 5, 8, 17, 18, 20, 21, 27, 29, 31, 32, 33, 34, 35, 37: Pixabay and 16, 17, 19, 25, 28, 39: Unsplash

Icons

Icons on page 12 were made by Alfredo Hernandez, Freepik, Nhor Phai and srip from www.flaticon.com.

Illustrations

Don Hatcher: pages 14 & 22. Simon Kneebone: pages 1 & 30. Angelo Madrid: pages 7 & 24.

Additional acknowledgements

With thanks to the Independence team: Shelley Baldry, Danielle Lobban, Jackie Staines and Jan Sunderland.

Tracy Biram

Cambridge, May 2019